George H.B. Macleod

Notes on the Surgery of the War in the Crimea

With remarks on the treatment of gunshot wounds

George H.B. Macleod

Notes on the Surgery of the War in the Crimea
With remarks on the treatment of gunshot wounds

ISBN/EAN: 9783337015688

Printed in Europe, USA, Canada, Australia, Japan

Cover: Foto ©berggeist007 / pixelio.de

More available books at **www.hansebooks.com**

NOTES ON THE SURGERY

OF THE

WAR IN THE CRIMEA,

WITH REMARKS ON

THE TREATMENT OF GUNSHOT WOUNDS.

BY

GEORGE H. B. MACLEOD, M.D., F.R.C.S.,

FORMERLY SURGEON TO THE CIVIL HOSPITAL AT SMYRNA, AND TO THE GENERAL HOSPITAL
IN CAMP BEFORE SEBASTOPOL;
LECTURER ON MILITARY SURGERY IN ANDERSON'S UNIVERSITY, GLASGOW, ETC.

PHILADELPHIA:
J. B. LIPPINCOTT & CO.
LONDON: JOHN CHURCHILL.
1862.

Dedicated

TO

SIR JOHN HALL, K.C.B., M.D., F.R.C.S.,

INSPECTOR-GENERAL OF HOSPITALS, AND PRINCIPAL MEDICAL OFFICER
OF THE **ARMY** WHICH SERVED IN THE CRIMEA,
ETC. ETC. ETC.

DEAR SIR:—

The permission to dedicate this book to **you, is** to me peculiarly gratifying.

I am glad to avail myself of **the** opportunity thus **afforded,** both of bearing my humble testimony, **as** a civilian, **to the un-**wearied assiduity and admirable skill displayed in the Crimea by **the Medical Staff, of which** you were the distinguished **chief; and also of** expressing **my** deep sense of the considerate kindness which **I have at all times** received from yourself **personally, more** especially **while attached under your command to the** General **Hospital in Camp.**

I have the honor to remain,

Your faithful Servant,

GEORGE H. B. MACLEOD.

ST. VINCENT STREET, GLASGOW,
May, 1858.

PREFACE.

No account of the surgical results of the war in the Crimea has as yet appeared, the only attempt to supply the desideratum being, so far as the author knows, some papers contributed by himself in 1855–56 to the *Edinburgh Medical Journal*. These being written hurriedly from camp, were of course unrevised by the writer when going through the press, hence the existence in them of many errors; while some of their statistics, although carefully compiled by the authorities in camp, have been found, by comparison at home, to be inaccurate. However, as these statistics were supposed to represent the results obtained during the entire war, while those now published only refer to the latter half, it is not improbable that they are more nearly correct than they thus appear to be. It is, therefore, thought that the following outline may not be unacceptable to the profession. It was printed several months ago, but its publication was delayed, in order to obtain the obvious advantages afforded by the government statistics. These, so long expected, have only just been finished, and

1*   (v)

a *resumé* of them will be found in the **Appendix to** this volume.

The author, **though for many reasons** regretting **the delay** which has taken place in submitting the following pages to the public, is gratified **however** to find that his views and deductions have been so completely confirmed by the Government Report, and by the work of M. Scrive, both of which have appeared since his book has been printed.

It is with great deference that this little work **is** given to the public; but the writer does so under the conviction **that some record of our** surgical experience **in the East** is desirable, and he bases his own claim to a fair and impartial hearing, simply on his having **had the good fortune** to see so much of the surgery **of** the war, **first at Constantinople** and Scutari during the greater part **of the** early period when the patients were chiefly treated there, and, latterly, **in** the Crimea during the last year of the campaign, when few cases left camp unconcluded.

The Crimean war, with its hardships and triumphs, has **passed into the calm page of** history, and remains only in **its** stirring memories **to those who** took part in it. Never, **perhaps, did any campaign** attract so great a share **of the** world's attention, or engage so much of its sympathies; and never again, perhaps, will such a concurrence of political circumstances bring together as friends or foes **so many nationalities.** England enlisted with heart and hand in the contest, and, it is to be hoped, has gained much from the experience it brought, as her future will, in no small degree, depend on the use that is made of this dear-bought experience.

But this great war has, unfortunately, added little to our medical knowledge. Its short duration prevented this; yet it has shown us wounds of a severity, perhaps, never before equaled; it has enabled us to observe the effects of missiles introduced for the first time into warfare; it has afforded us an opportunity of watching how dyscrasial disease may complicate injuries, and render skill abortive; and it has helped us to observe the development of those "diseases of circumstances" which may sweep away an army without any other weapon. Besides, every such war must furnish some surgical facts which are worthy of being chronicled, and must afford the surgeon some lessons which, without adding to his knowledge much that is absolutely new, are yet worthy of being remembered. A great war, in short, is a great epoch in the onward march of surgical science, when the slowly elaborated teachings of civil life are tested on a grand scale in the presence of representatives from every school.

If attentively sought, and carefully systematized, the experiences which are obtained in the field might become the most reliable and useful which can be anywhere collected, as nowhere are the circumstances which modify results more **easily** traced, or more uniform in their influence. Unfortunately, however, the vicissitudes, hardships, and uncer-

tainties of a campaign present difficulties of no small
moment to the collection and arrangement of observations.
The numerous duties which devolve on the military surgeon
prevent that close attention being paid to purely professional questions which would be requisite for the establishment of accurate conclusions, while the constant shifting of
patients from one hospital station to another occasions the
loss or interruption of records bearing on the treatment of
disease and injury.

The shortness and abrupt termination of the war were
unquestionably a great loss to the advancement of surgery,
however great was the gain otherwise to humanity. We
had, at its close, just overcome the preliminary difficulties
to be anticipated by a State long at peace, and whose military organization was defective; and when we might reasonably have expected, had there been another campaign,
to have garnered something valuable from its very miseries
and sufferings. But just when in a position to investigate
many questions with precision and advantage, the opportunity passed **away.**

The value to **be attached to** the statistics of any war
must be left to the reader. The writer believes that those
having reference to that in the Crimea are as correct as
any can be which are collected under such circumstances.
All the figures given in the body of the book, except when
otherwise stated, refer merely to the period after April 1,
1855, as it was found impossible to compile them with any

accuracy for the previous period. In this way a very large number of the wounded, and very many operations, are not **included** in these figures; and hence, too, why a different bearing must be given to many of the questions discussed than these figures will warrant. Thus, for example, the writer has himself seen more cases of some operations performed in the East than appear in the returns. It was thought better, therefore, to restrict the statistical enumeration to the period whose records were correct—although it is always to be remembered that what was true of the latter **part** of the war is by no means substantiated by the experi**ence of** the first part.

The writer has to acknowledge **with** gratitude his obligations to **Dr.** Smith, Director-General, not only for supplying him with the figures contained in the **body** of this book, but also for affording him free access to the reports on the China **and Indian wars.**

To Deputy-Inspector Taylor, C.B., of Chatham, **and Professor Tholozan, of the Val de** Grace, now first physician **to** the Shah of Persia, the writer is under many and deep obligations, as well **as** to Professors Mounier and Legouest, **of** the Val de Grace, who kindly communicated the details of their service at Constantinople. The writer would also express his thanks to **the** many friends who have supplied him with notes of cases, **some of** which are given in the following pages as illustrations of the questions discussed.

The value of surgery is nowhere so appreciated as on the field of battle, and the author rejoices to acknowledge how nobly our art is represented by the present race of military surgeons. In guarding the health, treating the wounds and sicknesses of his fellow-soldiers, the surgeon must truly participate in their glories and triumphs. "En les arrachant **aux dangers de** leurs blessures," says the famous Percy, "leurs triomphes deviennent notre ouvrage; la vie qu-ils tiennent de nous, nous associe en quelque façon à leur gloire; et chaque service que reçoit d'eux la patrie, est un présent dont elle est encore redevable à nos soins."

CONTENTS.

(xi)

CHAPTER VI.

CHAPTER VII.

CHAPTER VIII.

CHAPTER IX.

CHAPTER X.

CHAPTER XI.

CHAPTER XII.

APPENDIX.

NOTES

SURGERY OF THE CRIMEAN WAR.

ETC. ETC.

CHAPTER I.

The History and Physical Characters of the Crimea—Its Climate and
Geology—The Changes of the Seasons during the Occupation of the
Allies—The Steppe-Lands of the Interior—Vegetation and Resources
of the Country—The Natives, and their **Diseases**.

As special reference is made in the following pages to
the diseases which prevailed in the British army during its
occupation of the Crimea, and to the marked influence
exercised by its climate on wounds, it may not be deemed
either irrelevant or uninteresting to make some preliminary
remarks, however brief and fragmentary, on the history and
physical character of the Crimea itself.

The Black Sea (Pontus Euxinus) and the countries which
bordered it were, in the ancient Greek and Roman mind,
more associated with all that was gloomy and horrible, than
any other portion of the world. Their poets shrouded
these regions in blackest darkness, and peopled them, like
Milton's chaos, with all "monstrous, all prodigious things;"
"Gorgons, and Hydras, and Chimeras dire;" or, at the
best, with monsters—in human form, but strangers to human
sympathies—ruthless and murderous, sacrificing, to a god-
dess as sanguinary as the Bowhanie of the East, every
unfortunate mariner that chanced to be cast on their rugged

2   13)

coast.* The Læstrigones, Cimmerians, and Tauri, by whom
the Chersonesus Taurica or modern Crimea was peopled,
were to the ancient mind the very type of all that was
savage and relentless, and the peculiarly bloody rites of
their worship have engaged the attention of poets, alike
ancient and modern.   Euripides, in his "Iphigenia in
Taurid," has immortalized the name of the cruel priestess
who presided over these horrid observances; and both
Goethe and Racine have transferred her evil fame to their
respective languages.   Æschylus has bound the tortured
Prometheus to some rock of this precipitous coast, and
opens his great tragedy with lines which convey to us the
distance and loneliness which he attached to the scene of
his hero's prison;† and the *Tristia* of Ovid has rendered
familiar to all readers of Latin, the dismay and despair
with which that poet regarded his fate when doomed to
dwell on the opposite shore.

It might be expected that a region associated with such
horrors and dangers, natural and supernatural, should form
the poet's favorite scene for adventures and achievements
transcending the ordinary experience and prowess of man.
Accordingly, the Crimea occupied of old the place given
to enchanted castles and dragon-guarded palaces in the
romances of the middle ages, where the most redoubted
heroes were to signalize superhuman bravery, faith, and
endurance.   Here the Argonauts triumphed over difficulties insurmountable to men of common mould; here Pylades
and Orestes met the crowning adventure of their arduous
course; and here it was, according to the opinion of many

---

* Virgil, Georg. lib. iii. v. 349, et. ss.
        Semper hyems, semper spirantis frigora Cauri.
        Tum Sol pallentes haud unquam discutit umbras.
    See also Ovid's Metam., lib. viii v. 788.

    † ΧΘΟΝΟΣ, μὲν ἐς τηλουρον ηκομεν πέδον
        Σχύθην ἐς οιμον ἄβροτον εἰς ἐρημίαν.

scholars, that Ulysses passed through the severest **ordeals** encountered during his many weary wanderings.*

But passing by poetic myth and legend, we find, even long before the historic period, some indications of the habits of the occupants of the Crimea in the rocks of its valleys, as legible to-day as they were thousands of years ago. The strange dwellings excavated in the chalk cliffs, and universally attributed to the Troglodytes, clearly prove that that people belonged to one of the great Scythian families, which at one period overspread the whole north-east of Asia and Europe. Then written history tells us, that after the Scythians, the adventurous and colonizing Greeks occupied the Crimea, carrying with them the enriching power of their commerce, and the refining influences of their civilization.

The flourishing establishments of the Milesians along the south coast (B.C. 500) and the powerful republic of Cherson on the southwest became the marts where the corn **of** the Crimean plains—then richly productive from careful cultivation—was exchanged for the luxuries and delicacies of Greece. But, more than this, costly furs, rare spices, curiously embroidered cloths, collected over a wide area, as well as immense quantities of fish, were exported from the harbors, and the art-treasures of Athens and of Corinth eagerly purchased in return.

The resistless progress of Rome toward universal conquest destroyed the territorial power of Greece, and necessarily weakened her colonies and dependencies, however distant. We accordingly find that the Greek cities of the Crimea were, shortly after this period, subjugated by **the great** Mithridates of Pontus, and deprived of their pre-eminence, which for many centuries they did **not** regain. Mithridates was, after a most heroic and long-continued struggle with Rome in the very **zenith** of her military great-

---

* **See** especially the able work of Dubois de Montpereux.

ness, at length overthrown, (B.C. 64;) and the Crimea, with
his other dominions, fell under the sway of the empire.

The country enjoyed consideration for a long time thereafter, as guarding the Euxine from the Sarmatians, and
other northern tribes, who eagerly sought its waters as the
highway to the rich and soft climes along its southern borders. At length, however, the Goths effected what had
defied the Sarmatians. Without dwelling on the romantic
story, which says that their great ancestor, Odin, fled from
the banks of Azof to escape the fetters of all-conquering
Rome, and on the shores of the Baltic vowed that he
would rear a race of warriors who should spoil the spoiler,
it is certain that, at the Christian era, the Goths were a
powerful people, occupying the modern Prussia and surrounding districts. Year after year, and age after age,
they fought with the Romans, always pushing southward
and eastward. In course of time, they advanced from the
Baltic as far to the southeast as the shores of the Euxine.
Many and fierce were their combats with the Romans, but,
though occasionally compelled to yield to the remnant of
disciplined valor still clinging to the old legionaries, their
dauntless bravery secured to them many victories. They
turned their course, however, from the Danube to the
Borysthenes, and after occupying the fertile plains of the
Ukraine for a time, they again sallied forth toward the
south. They speedily subdued the Crimea—the Tauric
Chersonese, or the kingdom of Bosphorus, as the peninsula
was then indifferently called—and thence quickly pushed
their way into the richest and fairest regions of the empire.
It is said that traces of Gothic features, and very faint traces
of the Gothic language, are still to be discerned among the
inhabitants of the Crimean mountains. They have, however, left no other trace of their occupancy; and for many
centuries nothing is known of the history of the country
except that it was repeatedly ravaged by various tribes who,

during the "great migration of nations," swept like successive waves from the northeast to the southwest.

At length, in the thirteenth century, a band of the Mongolian Tartars, known as the Khazars, fixed themselves permanently in it, and gave it the name of Crim-Tartary. The majority of its inhabitants at the present day are undoubtedly their descendants. During this period a bright gleam of prosperity—soon, however, to be again quenched in barbaric darkness—shone upon the Crimea.

The Genoese, through the fourteenth century, rapidly rising to importance from their commercial skill and maritime enterprise, perceived the importance of the country as a link of communication between Europe and Asia, and purchased permission from the Khans to establish mercantile factories on the coast. These speedily became flourishing marts of trade, where the goods of northern Europe, of Asia Minor, of Persia, and even of the distant Indies, were collected and exchanged. Kaffa, now Theodosia, originally a Milesian city, rose to great wealth and power, containing a population of more than 100,000, and in many respects vying with the proud Italian city herself. These merchant princes have passed away; but even now, along the sounding shores, and far amid the mountains, in lonely valleys and on lofty hills, there remain many traces of this remark-**able people in** the nodding towers of crumbling fortresses, and the massive fragments of more enduring and noble architecture which have stood through many ages in solitary desolation.

Kaffa continued to grow and to prosper for about a century and a half. At the close of that period it suffered the doom inflicted on greater and more powerful cities **by a** ruthless destroyer. Mahomet II, the Turkish sultan, as is well known completed the destruction of the eastern empire by the capture and desolation of Constantinople in 1553. About ten years thereafter, he overthrew the noble Trebizond, and, still unsated, he in the year 1575 subjected Kaffa

to the same melancholy fate. He spared the buildings, but
carried more than 40,000 of the inhabitants to Constantinople in order to repeople its wastes, and took with him
many shiploads of gold and silver, and the richest merchan-
dise. The remaining inhabitants were scattered or destroyed. Kaffa soon became an absolute desert, and the
other Genoese cities speedily shared its decay.

Mahomet perceived his error, and sought by means of
Venetian settlers, to whom he offered the highest commercial privileges, to restore the prosperity which he had so
wantonly destroyed; but all history shows that even the
most powerful despot cannot restore a city once brought to
ruins, and the experience of Mahomet powerfully confirms
the general truth. Even Constantinople did not revive, and
Kaffa utterly perished in his hands.

For another period of about two hundred years the
Crimea languished under Turkish rule. About the year
1760, Catherine II. of Russia, seeing its importance as an
outwork whence to push conquest, annexed it to the Russian empire; and that her policy has not been forgotten by
her successors, has been too fearfully manifested to all Europe
in the late great struggle.

In the late war modern Europe has for the first time made
the Crimean plains the battle-ground, and I doubt not that
the acts of heroism to which they have been witness, the
patient endurance and stern bravery which have been there
displayed, will print the name of that land more indelibly
on the page **of** history than could **all the** transient glories
of rival cities in days long past, or the barbaric splendor of
the many conquerors who have successively ruled and passed
away from it forever.   At some distant day, when the
memory alone remains of the brave deeds which were there
performed, the husbandman will perhaps turn with the same
astonishment to gaze on the rusty and broken weapon, or
marble fragment, with, to him, unknown inscription, which
he has laid bare with his plow, as we did when we drove our

approaches over mosaic pavements, and by long-buried hearths, and deep among graceful capitals and shafts of shining marble, which alone remain to tell of the luxury and magnificence of departed nations.*

The Crimea is a peninsula of a quadrilateral shape, having a superficies of between 10,000 and 11,000 square miles, lying between the latitudes 43° 40′ and 45° 40′ N., and in E. long. 34° 30′ to 35° 30′. It is surrounded on all sides by water. The Black Sea washes its shore on three sides, while the Sivashe, or Putrid Sea, and the Sea of Azof complete its boundaries. Although connected to a great continent by a very narrow isthmus, and encircled by water on all sides, yet its near neighborhood to the mainland prevents its having a purely insular climate. It is by this peculiar position that we explain the great variation of the climate. The oscillations which take place between the two great climatic types—the continental and insular—impress themselves strongly on the attention, and are extremely difficult to reduce to fixed principles.

The southern coast, bold, steep, and inhospitable, has been torn and split by volcanic action, and indented in various places into deep and narrow harbors. The great force of that volcanic upheaving which elevated this coast is distinctly evidenced all along the shore. One great wave of burning rock extends from east to west for a hundred miles, lying close to the sea, and attaining an elevation of from 800 to 1000 feet. From its elevated crest the country slopes away north, in green grassy plains, which gradually melt into the wide expanse of the steppes.

At Balaklava, conglomerate, mixed with coarse sandstone and variegated marble, lies heaped up in wild confusion, forming vast masses which overhang the harbor. East

---

* The French established their approaches and batteries against the Quarantine bastion through the ruins of the ancient city of Kerson.

of this point, rocks, formed chiefly of tertiary limestone and colored marbles, throw their vast bulk many hundred feet into the air, or lie, like the Aia-dagh, or "Bear Mountain," in huge detached masses far in the sea. Jets of porphyry are often seen to fill the rents in the perpendicular face of these rocks, and at intervals the long dormant craters of extinct volcanoes are met with. To the westward of Balaklava, toward Cape Kersonesus, basalt, amygdaloid, and porphyry are seen under the tertiary, or steppe limestone of the plateau next the sea.

The calcareous composition of most of the mountains along the coast is plainly betrayed by their rounded summits. The sub-lying rock of the inland plains is limestone, chalk, and green sand, while in the valleys especially, the horizontal beds of these latter strata are particularly conspicuous. **The** precious metals are said to have been at one time found in the rocks of **the coast; but recent** investigation has failed to discover them.*

The appearance of the country some short distance from the sea is very curious and striking. Bold promontories, which look like sea capes left by the tide, occur in frequent repetition, standing amid encircling valleys, with their perpendicular faces always to the south, and sloping gradually to the north. Such inland promontories, from the ease with which they can be isolated and defended, have served **at all periods** of the country's history as natural fortresses for the inhabitants, and have proved, as in the case of the famous "Mangoup Kali," almost impregnable encampments.

The portion of the Crimea on which stand the "bloodstained" ruins of Sebastopol, and on which the allies were so long encamped, forms the lesser peninsula which projects from the greater at its southwest extremity. It is thus the lesser Chersonesus, or the Chersonesus Heracleotica of the ancients, deriving its name from Heraclea, the native city of

---

* See Appendix A.

the colonists, who built Kerson on its extremity.   It is
bounded by the sea, the harbor of Sebastopol, the Tchernaya River, which empties itself into the head of that arm
of the sea, and finally by a deep and broad valley which
runs across the neck of the peninsula from the Tchernaya
to Balaklava.   This piece of country, which is of a trian-
gular shape, measures some eleven miles from apex to base,
and about nine miles in breadth, containing an area of
about sixty square miles.   It was at one time divided from
the mainland by a wall built by the ancient colonists to protect themselves on the land side, and which barrier ran
across the valley from the River Tchernaya to the harbor of
Balaklava.   Within this boundary an enormous city once
stood, containing as many inhabitants as the British army
which more lately pitched their tents on the site of its ruins.
From Cape Kersonesus, the apex of the triangle, the land
of this little peninsula rises inland till it reaches some high
cliffs of fossiliferous limestone which immediately overhang
the valley, and the river which, I have said, constitutes its
boundaries.

The mountain chain which, I stated above, stretches along
the coast, strikes inland when it reaches Balaklava, and runs
for some distance in a northerly direction, then sweeps round
to the north side of the grand harbor of Sebastopol, and
dips into the sea.   This mountain barrier is very precipitous
on the side facing the plateau, presenting, below M'Kenzie's
farm, an almost perpendicular face of chalk of 1000 feet to
the summit.   This range is penetrated by various deep valleys, some of which give passage to copious streams.   It
will be thus understood that "the plateau of Sebastopol,"
as it was often called, upon which the British and French
troops were encamped, and which was the scene of so much
suffering and heroism, was totally excluded from the shelter
of the mountains, which, in fact, did not approach within
miles, but were sufficiently elevated above it, and adequately
near, to allow the enemy who held them to overlook our

whole position—an advantage which he turned to good
account. From these circumstances, the unprotected position of the camp to the north and east, and its complete
exposure to the sea on the west and south, will become at
once evident. It is of consequence to keep this in mind, as
it had a material bearing on the climate.

The elevation of the plateau above the sea level was not
great, being but 700 or 800 feet at its highest point. The
soil was generally scanty and light, but here and there it
consisted of a stiff clay. It was easily converted by rain
into a most tenacious mud, which interfered greatly with
progression, and was sufficiently adhesive to wrench the
shoes off the horses' feet. The underlying rock was, for
the most part, a porous stratified limestone dipping westward, and underlaid toward Inkerman **by** nummulitic limestone. From the lie of some of the deep non-porous strata
the water was conducted along them into the deeper inden-
tations and valleys which mark the surface of the plateau,
and there especially the **soil was** generally kept damp and
tenacious. **Along the sides of the** numerous ravines which
broke the surface in their course toward the sea, the rocks
cropped out in rough masses, and strewed the hollows with
detached fragments.

The valley on which the harbor of Balaklava opens, and
which I described above as crossing the base of the plateau,
consists of schists on the side next Balaklava, and farther
on, toward the Tchernaya, of limestone. This valley is
about four miles and a half long by two broad, and is
divided from the valley of the Tchernaya by the Fedoukine
heights—chalky elevations of some 400 or 500 feet, on
which one division of the French army was stationed, and
along the base of which the battle of the Tchernaya was
fought. The river, which is not of any size, flows close
below these heights over a bed of marl and pebbles. Beyond
it, a plain, averaging about three miles in breadth, stretches
to the foot of the M'Kenzie heights.

The line which marks the mean temperature of the Crimea in January corresponds to that of Iceland, (32°,) while its July line bisects Madrid, (72°.) The mean for the whole year corresponds to that of the Isle of Wight, (about 50°,) which is five degrees farther north in latitude.

There are in the Crimea two regions which possess very distinct and dissimilar climates. I refer to the narrow belt of sea-coast which is inclosed on its northern side by the **Tauric chain,** and of which I made mention before, and the **much** greater division of the peninsula which, excluded from the embrace of these mountains, lies totally unsheltered from the cold winds that blow so unrelentingly at certain seasons. In this latter division lies the plateau of Sebastopol.

About 100 miles of coast—nowhere of any considerable breadth, but at many points a mere stripe—is thoroughly protected by the mountain range which borders it. This chain throws the dreaded north and east winds which desolate the inland plains far beyond the limited belt, and secures for it a purely insular climate, little varying, and of the most delicious mildness. This region is the Baiæ of Russia, and contains the magnificent summer residences of her nobility. At Balaklava the mountains cease to follow the **coast, but** throwing one spur into the sea run north toward **the interior, where I** before traced them. It is this spur which so completely closes in these Elysian fields to the Sebastopol side, and which also enabled the enemy, from the difficulty of the passes, to maintain their ground, while the allied armies were kept outside on the bare plateau, exposed to the unmitigated fury of the winter's storm. Thus it was, that within a few miles as great a difference of climate was found as commonly exists between the opposite sides of a great continent, and while the outposts of the enemy's forces which occupied the southern littoral enjoyed during winter all the luxury of an Italian sky and tropical vegetation, our army lay in the sweep of those northern and

eastern blasts which, blowing over the vast frozen plains of
the interior, bring with them the rigor of an Arctic winter.

I have already hinted at the great variableness of the
climate of that part of the Crimea occupied by the allied
forces. This inconstancy was undoubtedly its most striking
feature. The mean temperature of a week or a month might
not, indeed, differ greatly from that of the week or the month
which preceded or followed it, but the daily sensible variations were frequently very great.* At some seasons these
diurnal changes were more severe than at others. Thus, in
winter, something of the following succession not uncommonly occurred : A dark, "muggy morning," with, perhaps,
a " Black Sea fog" rolling its heavy, damp folds over the
plain, giving you, when inclosed in its embrace, a feeling that
resembled nothing so much as being in the drying-room of
some large wash-house, would be succeeded by a splashing
rain, a sharp hail-storm, and an intense frost, all within a few
hours. The alternation of frost and thaw was sometimes
very remarkable, and it was difficult to tell which would ultimately prevail. A heavy fall of snow occurring during the
night would have its surface rendered crisp and dusty by the
keen morning frost, and at mid-day be converted into a deep
slush by the hot, sultry breath of the sea wind. Even so it
must have been in those dark and stormy days when Ovid
looked out from his place of exile on the wintry sea, and
wrote his dreary account of a Pontic winter.

Successive seasons appear, from the record of travelers,
to differ considerably the one from the other. Cycles, too,
of a similar character, alternate with others of a different
description. During the stay of the allies, the first winter
was providentially much less severe than the second, when
the preparations to meet it were more complete. A dreadful
severity appears periodically to mark the cold season. Fearful snow-storms, greatly dreaded by the inhabitants, pass like

* See Appendix B.

whirlwinds of death and destruction over the exposed parts
of the country, and bury whole villages beneath their drifting
eddies, while those icy winds, which few who have ever felt
their edge will forget, lash the unfortunate traveler as with
stripes of scorpions.

The transition from winter to spring is very rapid. A few
days revolutionize the year. In 1856 this was markedly the
case. To winter's "ruffian blasts" quickly succeeded—

> "Those softer gales, at whose kind touch
> Dissolving snows in livid torrents lost,
> And mountains lift their green heads to the sky."

Then, for a time, the climate became delicious. All nature
was in a moment astir, and awoke suddenly from the long
winter's sleep. We, too, "felt the spring in all our pulses."
Music was again heard in the various camps, which became
the scene of vigorous sports and healthful labor. Even
through the thickly-scattered fragments of the deadly shell,
the spring flowers timidly pushed their gentle heads; while
down the little valleys, and by the water-courses, the crocus
and snowdrop, with various orchidaceous plants, reminded
one of similar sheltered nooks at home.

Summer, again, is for the most part oppressively hot. In
June, July, and August the temperature ranged from **80°**
to 100° on the plateau, and this, during our occupation, was
often far more than could be borne with comfort, protected
as we were only by the thin covering provided by cam-
paigning resources. It was at this season that the hated
sirocco tainted the air with its hot, oppressive breath, which,
while it turned the strength of the healthiest into feebleness,
utterly prostrated the poor invalid. Then, too, the nauseous
effluvia of a great camp became most obnoxious, and the
loaded air refused to the sick and languid that refreshment
they so greatly needed. The dryness of the wind which so
often prevailed at this season removed the moisture from
the body as soon as it was formed, and produced an amount

3

of lassitude difficult to describe. If the nights had been fresh, the effects produced on the sick by this weather would have been less destructive than they were, but, unfortunately, the breeze often died away at sunset, and rendered the evening hours very exhausting. After the unfortunate assault of the 18th of June, this circumstance was very marked, and proved highly injurious to the wounded. Dew appears but little in summer—at least so it was during our stay. Thunder-storms occurred occasionally, preceded by whirlwinds, which carried columns of dust through the camp, and filled every corner with **their burden.** At times, tents and huts went down **before the suddenness** of the assault, stifling and crushing **the** helpless sufferers who lay within them.

Autumn again was charming. The coolness of the mornings and evenings, the tempered heat of the mid-day sun, and the genial showers rendered the climate delightful, while the beauties of sun and shade which played in such variegated richness over the broken mountains, and the brilliant sunsets which lit up the many peaks with their purple splendor, and threw their golden shafts far over **the** shining surface of the calm sea, gave camp-life, at that season, a charm to which we had previously been entire strangers. Notwithstanding the many predictions to the contrary, the health of the troops did not, at that season, deteriorate ; the " Periculosior æstas autumnus longe periculosissimus," though loudly proclaimed at home, happily found with us no verification.

Taking one season with another, the climate of the Crimea must be admitted to be a fair one, especially when the inhabitants are protected by well-built houses. That our army suffered from very few ailments which were directly referable to the climate, says much for its goodness, when it is considered how little it was protected against the rigors and **vicissitudes of** such weather as prevailed. The officers, although exposed alike with their men, yet suffered far less from sickness—a result which can be attributed alone to the better diet, the change of clothes, the superior bedding, and freedom from manual work, which they enjoyed. The extraordinary immunity from pulmonary affections which existed was, at least, remarkable; yet, that certain seasons did very decidedly impress their influence on the wounded, I am ready to maintain. Of this, however, I will speak more fully afterward.

The steppe lands of the interior present, in early spring, a lovely scene of rolling plains of waving grass and gay flowers, **stretching** in unbroken verdure to the horizon. As the heat increases, all this vegetation, however, is soon burned up, from the soil being but thinly strewn over the underlying tertiary limestone. At midsummer, the view is dreary and cheerless enough. Clouds of dust, carried by the whirling currents of air, sweep over the wide, shadeless expanse, and the trembling mirage wanders restlessly, or the " burian " wheels in weird-like flight before the burning breeze. **No** mountains relieve the eye, and only the swampy salt lakes, which occur at intervals, break the dead uniformity of the barren waste. An oppressive monotony reigns everywhere. The brown, changeless plain and the brazen sky overhead continue ever the same; not a shadow is cast across the horizon, and day sinks into night almost without a twilight.

Many of the valleys among the mountains of **the sea-**coast form, on **the other** hand, a delightful contrast to the dull, dreary waste of the steppes, being very beautiful, highly cultivated, and well peopled. The fields were literally covered with fruit-trees when I first saw them in spring, and the neat, clean villages of the Tartar peasants stood clustered round with many-colored blossoms. Clear, sparkling streams from the overhanging mountains watered the **green** pastures, which looked like nothing so much as a constant succession of gardens.

The vegetation of the Crimea **differs** much in different parts. Along the sheltered south coast it is almost tropical. The vine there grows in luxuriance, and yields a wholesome wine. The mulberry, the fig, and the olive, the pomegranate, filbert, and walnut, together with the peach, the apple, the pear, apricot, and cherry, crown the hillsides as with a forest; while rare flowers flourish without protection in the open air. In other parts of the peninsula, elm, ash, and beech, together with the juniper and pine, are met with; and high up, clinging to the precipitous rocks, the Tauric pine retains its unstable footing. The wild rose, asphodel, iris, primrose, and hyacinth deck the ground at various seasons, and the peony astonishes our English notions by its size and fragrance.

Game abounds in many parts of the peninsula. The hare and red-legged partridge, with wild duck and snipe, were occasional and welcome guests within our lines. In winter, the greater and lesser bustard and the wild goose passed in vast flocks over the camp, and were sometimes initiated in the mysteries of the *pot-au-feu* by their fortunate captors.

The present natives of the country are of a mixed description. Those of the plains are undoubted Tartars, with the high cheek-bones, far-severed pig-like eyes, and flat features of the Mongolian; but the inhabitants of the mountain **villages** appear of a much more mixed race, and, not uncommonly, the refined features of the Greek physiognomy are met with. They appear a quiet inoffensive race, honest in expression of face, and powerful in body. They are very cleanly in their persons and habits, and their villages are **generally models of neatness. Their** houses, collected into little knots in the villages, or perched in the sheltered nooks of the mountains, though unpretending, have an air of comfort not often seen in the East; and the peaceful demeanor of the inmates, who crowd out to stare at the foreigner, seems to belie their ancient character for warlike **prowess** and **ferocity.**

I could learn but little of the diseases of the inhabitants, from not knowing their language. Scabies seemed not uncommon, and many were pitted with small-pox. Deputy-

Inspector Mowat, in his report on the Russian hospitals, tells us, on the authority of the Russian medical officers, that remittent fevers appear to be the endemic of the Crimea, and that disease from which the Tartar inhabitants chiefly suffer. Ophthalmia is **said** to be common in Sebastopol, caused probably by the fine limestone dust of the unpaved streets, and the strong sunlight. Though having my attention fully alive to it, I failed to see any cases of tubercular elephantiasis, **the** "Morbus Crimensis" of Pallas and Martius, and I never heard any one say he had seen a case in any of the villages. Probably, if such cases did exist, they would **be invisible** to the traveler, as the inhabitants are too well off to show their sick as objects of charity.

3*

CHAPTER II.

I HAVE endeavored, in the previous chapter, to describe
the physical conformation of the Crimean peninsula, and the
leading features of the plateau on which were established
the cantonments of the allied army. When this plateau was
first occupied, a thick brushwood covered a considerable part
of its surface toward Inkerman. Most of this underwood
consisted of low bushes growing out of the "stools" of former trees, and was seldom of any considerable size. It is a
well-recognized fact that ground so covered, or which has
been but lately cleared, forms a most objectionable position
for a camp, chiefly on account of the moisture which is so
apt to be retained by a soil shaded from the sun; but in the
present instance the evils which might have followed the
enforcement of strategic expediency in preference to the dictates of medical prevision, were providentially counterbalanced by the elevation, height, and openness of the ground,
and the absence of trees of any magnitude.

The low ridges and rounded knolls which occurred in
frequent repetition over the plateau afforded most excellent
sites for tents and huts. By guarding them with a trench on
the uphill side, the surface water was diverted, and the
ground which they covered kept dry. Advantage was taken
of the elevations to increase the accommodation and comfort of the tents by digging a hole into their sides, over
which the tent was pitched. During heavy rains these holes

were, however, sometimes filled with water by the overflow of a deficient drain, and fatal consequences have followed the use of charcoal fires in them, from the weight of the **gas** generated keeping the pit full of a dangerous atmosphere.* The chalky subsoil of the plateau was very conducive to the health **of the** camp, both from the rapidity with which the surface dried, and also from the springs which it afforded. As, however, the water flowed along the surface of the deeper non-permeable strata, and welled out at those points **where the** ground sank below their level, the valleys and deeper depressions were usually boggy, especially at their heads.

The dip of the plateau and the run of the numerous ravines being toward the coast, the **drainage** of the camp was everything which could **be** desired. The hollow below the guards' camp, the piece of ground situated near the headquarters of General Bosquet, and behind the sutlers' village, known in camp parlance as "Donnybrook," and a **plot** of land under Cathcart's hill, were the most objectionable **parts** of the encampment. These spots **were** soon **avoided**, as their pernicious effect on the health of the troops sent to occupy them was soon evident. Marshy and unhealthy ground existed along the Tchernaya, especially near its mouth, and at the head of several of the creeks; **but most of these points lay** beyond the **allied lines.** The French division, which was at a late period of the war cantoned along the river, suffered considerably from the marsh miasma, within the range of whose influence they there entered.

The **water** obtained within our lines was, on the whole,

---

* The Russians, Sardinians, and Turks constructed **underground** huts for winter **use; but the difficulty** of **ventilating, lighting,** and **keeping** them dry more **than** counterbalanced **their** advantages as to **heat.** Malignant fever **has** been said **to** increase much in its virulence among troops so housed. How **much it** may have contributed to cause and maintain it among the Russians, I am not in a position to determine.

good in quality; but its amount necessitated economy in
its use during summer. The fear that the supply would
fail in autumn was one of the many evil anticipations which
the establishment of the siege gave rise to at home; but by
the prudent precautions adopted by the authorities in con-
structing tanks, and placing the wells under supervision,
these anticipations were fortunately not realized. Without
much attention, however, to the management of the reser-
voirs, I doubt not a scarcity might have been felt after the
exhausting drought of the hot summer and the increasing
demands of an augmented army. During midsummer the
water obtained within the position of the division to which
I belonged—the third—contained a vast number of animal-
culæ, many of them so large as to be seen by the naked eye
swimming about in little shoals in the water tanks. How
far the unavoidable use of this water may have predisposed
to that outbreak of cholera which took place at that season,
it is difficult to say. The wells were generally situated at
the head of the ravines, down several of which small brooks
flowed constantly.

The pressure of other more weighty matters at the early
part of the siege, and the unanticipated sojourn of the
troops on the ground, prevented that care being at first
taken in the arrangement of *latrines, slaughter-houses*, etc.,
which is so desirable in the distribution of a camp, and
which was afterward so abundantly shown. It appeared to
me that the arrangements on this head were better within
the British than in the French lines. The difficulty of
organizing this part of a camp is by no means trifling, and
the necessity of paying great attention to it is evident when
troops remain long stationary.* Pits were used as latrines

---

* Those who visited the encampments of the Russians on the
M'Kenzie plateau after peace, will easily be able to understand what
a miserable condition cantonments may be reduced to by the neglect
of the necessary precautions on this head. So to allocate sites for

by both armies, and earth or lime was thrown on the surface of the ordure. Generally speaking, the pits were too broad, as they exposed too great a surface to the atmosphere. In winter, the smell from these pits was scarcely perceived; but, during the heats of summer, and amid a teeming camp, it became much more obvious, and must have had a sensible though little recognized effect on the health of the troops, more especially on the progress of wounds.

The burning of the horse manure within our lines was a great mistake. The black fetid smoke hung in heavy folds over the camp, and was carried far and wide. The French with more wisdom buried it. Such trivial circumstances assume an unlooked-for magnitude when repeated on so vast a scale.

There was yet another point in the hygiene of the camp which did not at first obtain sufficient attention. I refer to the burying of dead carcasses. Many animals, horses, buffaloes, and bullocks died in passing to the front, or after arriving in camp, and from the want of men to bury them were not unfrequently left to decay where they fell. The stench from this source between Balaklava and the headquarters' camp was at one time very offensive, and at last compelled active interference for its removal.

During the latter part of our occupation, the arrangements with regard to latrines, shambles, the burying of dead carcasses, etc. were unexceptionable, and every means was taken to abate their baneful influence.

The food provided for the army during the first winter and spring was defective both in quantity and quality. This arose partly from unavoidable circumstances, and partly from inexperience in the officers to whose care was intrusted

these purposes as that they may be accessible to the camp, and at the same time sufficiently removed to prevent their contaminating the air, and that the locality they occupy shall not be required at any future period for the location of troops, calls for considerable forethought and arrangement.

the supply of the army. Salt meat and biscuit constituted
the bulk of the distribution, while rice, coffee, and sugar
were occasionally, but sparingly added. Sir Alexander
Tulloch says that, during December, January, and February,
"there was almost a total absence of fresh meat, and even
the sick were for many days, nay even for weeks, fed exclusively on salt meat, in their state a poison." The coffee
being served out raw and unground, was all but useless, and
the ration salt pork was not always of the best.

The want of fuel, and the state of fatigue in which the
men returned from duty, made them frequently eat their
pork half dressed, or toasted only before their meager fires;
and this, together with their ration of spirit, or it might be
their biscuits **and rum alone, formed their** frequent if not
their only fare.

This circumstance, taken into consideration along with
the prevalence of ulceration of the intestines **which existed,**
assumes an additional interest when connected with **a fact**
mentioned by Dr. Rollo, and quoted by Sir George Ballingall, that in the **year 1789 the** 45th Regiment lost, within
a short time, in Granada, and during a healthy season, **a**
large number of men who were found to have ulcerated intestines, and, on inquiry, it was discovered that one chief
cause of the mortality was, that the common breakfast consisted of a glass of raw spirits, with a small slice of boiled
salt pork, the spirits being not unfrequently repeated during
the day. May **not a like fare have been the** cause, to some
extent, of a similar effect in the Crimea also?

I have little doubt, that if the precaution had been taken
to supply the troops every morning with hot coffee, as they
went on or returned from duty, which was a step strongly
recommended as a prophylactic **at Walcheren, much of our
mortality might** have been **avoided.   It can hardly be
doubted that this** could have been accomplished at the worst
of times **by a** little management, as there are few things
more **portable or more** easily prepared than coffee. The

Turks place great reliance on this beverage as a preservative against dysentery, and the French preferred its use in their army to the tea which we employed. If we were ordered to prescribe a dietary the best adapted to give rise to gastric irritation and dyscrasial disease, could we suggest one more potent than salt pork, hard biscuit, and raw rum?

Men severely worked, and constantly in a keen air, require to have their physical energies sustained by a liberal **supply** of such food as contains the largest amount of nourishing and staple ingredients; but in place of that, the supply to our troops, besides being irregular in amount, was insufficient for their support, and those constituents which were most calculated to provide for their necessities were reduced at the very time when they were most required. Thus, in November the ration of biscuit and that of rice were altogether stopped, "so that within one week the troops were, in most cases, deprived of nearly half a pound of the vegetable and farinaceous food so much required to counteract the salt meat diet, and this, too, when scurvy had made its appearance."*

The want of fresh bread and vegetables was a great and serious privation, particularly felt by the sick, and those whose gums were **tender from** scurvy. Preserved vegetables, even when procurable, as they were not till late **in the** war, are at best but bad substitutes for fresh esculents, and lime-juice did not form part of the distributions till the scurvy poison had fairly impregnated the systems of the men.† It is useless now to inquire why that store of lime-juice, which is proved to have lain at Balaklava during the **two** months when scurvy most prevailed, was not distributed to the longing troops. The fact can now only be deplored,

---

* See Appendix **C.**

† The French, toward **the end of the war,** established gardens within their position, particularly along **the** Tchernaya, where they cultivated vegetables. These would have been of the greatest importance if the campaign had continued.

but the fault seems to have been one of the commissariat, not of the medical department.

The fresh meat, which was no less acceptable than rare, was not by any means invariably good. The miserable cattle arrived in the Crimea, after the transit over a stormy sea, in no very favorable condition for the butcher. Baudens tells us the French soldiers characterized them as Pharaoh's lean kine, and that the use of their flesh gave rise to intestinal flux of greater or less severity.

From a consideration of all these circumstances combined, in regard to diet and cooking, we derive the explanation, in a great measure, of the prevalence of certain diseases afterward to be specified, and the fatal virulence of others.*

The sick as well as the healthy were exposed to the evils arising from the defective rationing which I have been reviewing, but the praiseworthy and urgent efforts of the medical officers were frequently rewarded by obtaining medical comforts of various descriptions for the hospitals. Even after they got food for their patients, the difficulty of preparing it suitably was a great and trying one. It was one, however, which their humanity and energy surmounted to a considerable extent.

In all these remarks, I beg to repeat that I allude only to the early period of the war, as, latterly, every luxury prevailed in our hospitals, and our army lived as I suppose no army has ever before fared in the annals of warfare. The

---

* It is much to be regretted that the difficulties of transport make it almost impossible to vary the food of the soldier in the field. The constant repetition of the same rations, the absolute uniformity in every item of food, is but too apt to occasion aversion, especially with those in whom disease is beginning to show itself. I can speak **from** personal experience as to the strong predisposition of this one cause in giving rise to the fever designated "Crimean," and I know of few things which had a more undoubted effect on the health of the troops.

change which took place in this respect has occasioned the handsome compliment from the French medical inspector, in his review of the campaign : "Quand on compare les conditions où se trouverent les Anglais au debut de la guerre qui les prenait au depourvu et celles où ils s'etaitent placés en 1856 on est forcé de reconnaitre la grandeur de la nation Britannique."

For fuel the army was chiefly dependent, early in the siege, on the underwood which covered part of the plateau, and afterward on the roots of these bushes. They had to dig for this "underground forest" often beneath snow, always among wet mud, after the more fatiguing duties of the day were over, and so it was that much time was thus lost before they could procure, and still more before they could ignite, these wet roots, which were their only resource. A cheerful fire came to be almost unknown; and I have heard many of the survivors say that **few** objects appeared so frequently to tantalize them in their dreams. **It was** always difficult to obtain even as much firing as served **the** bare necessities of the camp. This deficiency **was** severely felt during the inclemency of winter, **and** enhanced greatly the other hardships; for it is a very true remark, **that** "a sufficient supply of **firewood** during a campaign is one-half of a soldier's existence."

The deficiency of clothing, which **was** so much complained of during the early part of the war, was one of the most prolific sources of subsequent disease among the troops. The soldiers' kits having been left on board ship when the landing **was** effected at Old Fort, and not being delivered to them till long afterward, compelled the men to perform **much** of the trench work in tatters during the severity of a Crimean winter. Their shoes, originally bad, **were** in many cases totally destroyed before they had **been long used,** and their only suit of clothes was soon reduced to shreds. The Quartermaster-General tells us that "they had had the suit they wore in the voyage out to **the** Mediterranean, through

4

the service in Bulgaria, through the sea-voyage to the Crimea; they had worked in these coats in the trenches, and fought all through with them; they were perfectly threadbare, and in many instances did not exist." All this, too, was allowed to take place, and the men to be exposed to the wet of winter, and severe cold in the trenches, while "thousands of coats were lying unused, and tens of thousands of greatcoats, blankets, and rugs filled the Quartermaster-General's stores, or the harbor of Balaklava."*

The one only blanket which each soldier possessed afforded, even when dry, but a feeble protection against the cold of the tents, but as he generally carried the same blanket with him into the trenches, it was commonly dirty and soaked with water when he came to sleep in it at night. The sole-less boots were seldom removed when he lay down to sleep, so firmly adherent were they to the swollen feet. Such was the condition of

"The poor soldier that so richly fought,
      Whose rags sham'd gilded arms."

The uniform which, of all others, seemed best adapted for the Crimea, was that worn by our enemies. The long, warm gray coat, gathered in folds over the loins, the low, flat cap, and the wide half boot, within which the trowsers were tucked, formed a much better provision against the cold winds of winter and the deep tenacious mud, than the dress worn by any portion of the allied army. The coats of the Sardinians approached nearest in shape to those of the Russians, and the French gaiters, though most serviceable, were not equal to the half boots of their enemies.

The housing of the troops during the early part of the war was in keeping with their food and clothing. At first

---

* The loss of the *Prince* was one great cause of the deficiency of warm clothing. In her went down 53,000 woolen shirts, 17,000 drawers, 16,000 blankets, 2500 watch-coats, 25,700 socks, 3700 rugs.

the common bell tent was used; huts were afterward added. This tent measures 13 feet 8 inches in diameter and 10 feet in height. It contains about 512 cubic feet of **air, has** almost no means of ventilation when closed, and was yet made to accommodate fifteen men, who lay at night without bedding on the bare ground. It may be easily imagined how vitiated the atmosphere of these tents was in the morning, when they had been kept close all night by their inmates in order to make themselves warm.

In pitching the tents, far too little space was, at first, left between them. In many camps, the ropes, instead of **being** stretched to their full length, were greatly shortened, which, while it unduly crowded the tents, necessarily lessened their stability. This arrangement was most injudicious, and did much, I doubt not, to render the camp more injurious to health than it would otherwise have been.

The earth upon which tents are pitched undoubtedly absorbs much animal effluvia, and comes to give out unhealthy emanations, which remain in the upper part of **a tent, and** can be got quit of only by striking it altogether, or removing it **to** another spot.

The huts at first used for barracks were that known as the "Portsmouth hut," which measured 27×15 feet inside, 6 feet to the **eaves, and 12** to the ridge. Each contained about 3645 cubic feet of air, and, when occupied by twenty-five men, allowed about 146 cubic feet to each. The Chester huts, which came to camp at a later period, were larger and better constructed than the Portsmouth. The errors, however, **of** all the huts were the want of sufficient independent means of ventilation, and the mode in which they **were usually** erected. Sufficient care was not always taken **to prepare the ground** by draining and covering it with loose rubble, **before** laying **the flooring. Their walls, too, except** in the case of the paneled huts, were too thin.

The **duty** which fell to be performed by the army was extremely heavy during the whole period of the siege, and es-

pecially during the first winter, when the amount of trench
to be constructed was very great, and when, from the number
of sick, a double share of duty fell on the effective.

It is well known that the extent of "approach" at first
assigned to the British army was very disproportionate to
its relative numbers as compared with the French. The
whole right attack of the combined army was appropriated
to our forces, and thus they had to form those vast trenches
(the more extensive as they were distant from the enemy's
works) which were afterward consigned to our gallant allies,
and made the basis of the triumphant advance against the
key of the position—the Mamelon vert and the Malakoff.

But trench work did not comprise the whole duty, as, when
not so engaged, the carrying of water and the procuring of
fuel engrossed the few remaining hours, so that the leisure
enjoyed was but scant, and the opportunities few, for con-
structing any means of protection against the cold. During
December, January, and February, 1855–56, the term of
duty in the trenches was so frequent that it required superhuman exertion. Thus, by the returns, it is shown that out
of an effective strength of 11,367 in January, 5321 were
told off daily for duty. The routine of this duty is well
illustrated by the extracts given in the appendix (D) from
Sir Alexander Tulloch's pamphlet; and if for a moment it
is realized—if we consider that this dreadful ordeal had to be
undergone day after day, and week after week, without any
intermission but that brought by the invasion of heavy sickness or the hand of death, then may we perhaps estimate the
effect of such duty on the health and constitution of those
who survived. The very fact that, during peace, our soldiers,
either at home or in the garrison towns abroad, live a life of
ease and plenty, made such unaccustomed duties peculiarly
severe, and the effect the more certainly destructive. But it
would only suggest half the truth were we to suppose that
in the phrase "trench duty," the hard, bodily exertion of digging was alone included. It was the standing ankle deep in

mud, or snow, or frozen water while they worked; the want
of shelter, and the absence of a dry resting-place when exhausted; the *mental* depression produced by such spade and
pickaxe work; the danger which accompanied it from sudden
sortie, bounding round shot, and exploding shell; the total
absence of all comfort on returning to camp, and the cease-
less recurrence, without apparent results, of the same routine,
which rendered this "trench work" so truly what the soldiers
called it, "desperate." The evidence of the mental depression to which I refer will present itself in dread remembrance
to those who can recall the condition of the soldiers during
the first five months of the siege; and its influence on the
outbreak and fatality of disease, as well as on convalescence,
cannot, I believe, be exaggerated.

From all that has been said, then, on the housing, clothing,
food, and fuel provided for the soldiers, and the killing toil
exacted from them during that period of the war which had
most influence on their health and after-history, the sad picture may be formed of their position in the Crimea. Day
after day passed in severe bodily exertion and anxious
watching—one moment digging laboriously in extending the
approaches, and the next with arms in hand repelling the assaulting enemy; almost always wet; exposed without cover
to the drenching rain and soaking snow, the keen frost and
biting wind; standing for days in wet mud; constantly either
unnaturally excited or depressed; ever in danger and without hope of a change; their dirty, humid clothes in rags,
their bodies covered with loathsome vermin which seemed to
grow out of their very flesh; no comforts in their wind-
pierced tents on the bleak plateau; no fires, unless, weary
and foot-sore as they were, they dug beneath the **snow**-
covered sod for wet roots wherewith to **kindle a feeble** and
tantalizing blaze; without food till, after **hours of** perse-
vering exertion, they managed to half **cook their** unpalatable
ration over their winking fire; huddled into a crowded tent
to pass the night in a close, noisome atmosphere, on the oozy

ground, covered by the same blanket which protected them
in the wet and muddy trenches; longing for the morning,
though its early dawn was signaled by the bugle sound which
called them to a renewal of that dread task whose severity
made them yet again sigh, "would to God it were night."
This sketch is no exaggeration. It is true, though difficult to be realized even now by those who themselves saw it.

Can we, then, find anywhere else in reality—nay, can we,
by the utmost stretch of imagination, conceive a more fruitful
field for the seeds of disease, or the harvest of death, than is
here presented to us in the camp of the weary, anxious-
minded soldiers who fought so gallantly, endured so con-   .
stantly, or died so nobly, and who now consecrate, by their
humble graves, the green hillsides and lonely valleys of the
Crimea?

The bare remembrance of that frightful combination of
circumstances which seemed to encircle **our army as with**
ever-contracting walls of iron, and make it prisoners for
those dread scourges, cholera, fever, and dysentery, that,
like the angel of death in the camp of Sennacherib, **de-**
stroyed our noble and gallant army, comes to one's memory
like the awful vision of a distempered dream.

These things must be here recalled as they had a most
important influence on the annals of the war, and much that
would otherwise be unintelligible becomes clear as noonday
when read by the light thrown on it by these circumstances.*

---

* "The poor condemned English,
Like sacrifices, **by** their watchful fires
Sit patiently, and inly ruminate
The morning's danger; and their gesture sad,
Investing lank-lean cheeks, **and** war-worn coats,
Presented **them unto** the gazing moon
So many horrid ghosts."—KING HENRY V., act. iv. scene 1.

How **completely** does the noble heroism displayed by our troops
in the Crimea refute General Foy's estimate of them! "The Eng-

But if such a condition of things as I have feebly outlined was trying to the strong, how can I express its influence on the weak! It is impossible fully to realize the hopeless condition of the sufferers, struck down by enfeebling sickness or exhausting wounds, and deprived of that vigor which alone made hardship endurable.

The regimental hospital marquee and the round bell tent which served as hospitals were of necessity vastly overcrowded.* The former, which measures 27 × 14 feet inside, and affords about 3250 cubic feet of air, ought to have contained only twelve to fifteen men, but was made, from the exigency of the service, to cover three times that number. The unsuitableness of the bell tent to hospital purposes has been fully expressed by the commissioners when they say: "Whatever may be the supposed advantages which have led to its adoption as a barrack tent, it would be difficult to contrive anything much more unfit for the accommodation of the sick." There were no bedsteads, and those patients who had empty sacks to lie upon were considered fortunate. Few blankets belonged to the hospitals, and the food which

---

lish soldier," he says, "is not brave at times merely; he is so whenever he has eaten well, drunk well, and slept well. Yet their courage —rather instinctive than acquired—has need of solid nutriment; and no thoughts of glory will ever make them forget that they are hungry, or that their shoes are worn out."—Foy, vol. i. p. 231.

* If properly constructed and erected on suitable ground, there are no structures better adapted for the hospitals of an army in the field than wooden huts or canvas tents. The dreadful epidemics which have so frequently pursued armies, and the mortality which has attended their wounds, have in not a few instances been due to the employment of stone buildings as hospitals. The ventilation is more apt to be deficient or to become deranged in them than in huts or tents, and hence the effects of overcrowding become the more pernicious. "It was often proved in the history of the late war," says Jackson in his work on the economy of armies, "that more human life was destroyed by accumulating sick in low and ill-ventilated apartments, than in leaving them exposed in severe and inclement weather at the side of a hedge or common dike."

they afforded was but ill adapted for sick men—nay, in
many instances, constituted a veritable poison; medicines,
even the most necessary, but scantily provided; attendance
by overworked, and in many cases sick doctors, and **by a**
handful of orderlies, themselves for the most part convalescents, whose natures, however kind originally, must have
become soured and crabbed by the hardships and fatigue to
which they were exposed—such was the condition of the
sick during the first winter. If to sickness were added
wounds—a broken limb or contused body—how small was
the chance of recovery! Splints and bandages merely
teazed and fretted. The man lost hope. Every circumstance forbade recovery. The powers of evil seemed to
grasp his destiny. The problem of life was being solved,
by every conceivable antagonism having **a voice** in the
momentous decision. Such being the circumstances given,
how could any other result follow than a mortality which
caused our land to ring with the voice of mourning, and
which for a moment paralyzed **our** senate and our people?
The wildness of despair is the only excuse which can be
made for the blame of so much misery having been **cast on**
the medical department, which had no control whatever
over the events that led to it, and the voluntary sacrifice of
whose members, though glorious to themselves, was unable
to retrieve the deplorable errors committed by others.

During the early part of the war, the regimental hospital
system, which I believe is peculiar to the British army, was
alone **followed** *   At a later period, general hospitals were

* According to this plan, the soldiers are kept, when sick, in hospitals which belong to their regiments, in place of being transferred
to general hospitals established to receive the common sick of the
army. The discussion is an old one as to whether the system which
has been always followed by us, except in times of great necessity,
or that of general hospitals, the plan adopted by the French and
other continental nations during war, is the best. I do not intend
here to review the question, but would merely remark that, after

established. One large building at Balaklava, which had
been a military school, was early appropriated to this pur-

fairly weighing the subject, I think the regimental hospital arrangement presents the greatest advantages. One strong claim which it
has on our support is, that the surgeon of the corps must be greatly
better acquainted with his men, their character and habits, and thus
be more able to treat them, as well as more able to detect imposture
if attempted, than the medical attendant who probably never saw
them till they present themselves to him as patients. The men are
by this system also kept among their comrades—no small advantage
to them—and thus their minds are cheered by the companionship of
friends. More time and attendance, too, can be bestowed on the
sick in these small hospitals, from the proportion of medical attendants and orderlies being in general much above what it is in the
larger establishments, and from the responsibility being, if possible, more binding on the regimental surgeon, and the stock of comforts proportionally greater and more regularly supplied, from the
resources of the whole regiment being at his command. In regimental hospitals it is well known that wounds heal more satisfactorily, and that purulent infection and gangrene occur more rarely,
most probably because, from the mixture of cases which takes place
in them, that segregation of suppurating wounds which is so apt to
occur in general hospitals is avoided.

On the other hand, a much larger staff is required when the hospitals are confined to regiments. A regiment numbering one thousand men on active service has a surgeon and three assistants attached to it, all of whom are rendered useless if their corps is not
engaged, while a superabundance of work falls to the lot of those
medical officers whose regiments have suffered severely in action.
Hence it follows, that if the whole medical staff of the army was
united and concentrated in general hospitals, less than one-half of
the aggregate number of professional men would suffice for the service. In the French army, the proportion of medical officers to the
strength is not a third so large as in ours.

Besides, a more uniform system can be followed in the treatment
of the cases, and the results of such treatment made more available
both for instructing the younger surgeons, and also for the promotion of science, when conducted in large hospitals, than it can ever
be in small detached establishments. The cost of administration
also will be greatly diminished by concentration, and the whole

pose; and, as the number of sick increased, huts were
erected to add to its accommodation. This hospital was
chiefly used as a depot for the sick about to embark for the
Bosphorus, and for the treatment of sailors and native
laborers. The position of the hospital was most unfortunate. In summer, it was a perfect furnace, "perched as
it was in the focus of a concave mirror, of which the sides
were formed of bare rock, and the bottom by the smooth
water of the harbor;" and its near neighborhood to the
town was a great disadvantage.

Above Balaklava, on the face of the precipitous rocks of
the coast, a number of Portsmouth and double-walled huts
formed the *sanitarium*, to which convalescents were sent
from camp. The exposed position of this hospital made it

---

economy of the army, as regards the management and transport of
the sick, easier arranged and much more efficiently conducted. But
with all this—and it is a very great deal—to be said in favor of large
hospitals, yet I unhesitatingly think that, tried by the one great test
of the saving and prolongation of life, which in our army at least
is the chief criterion of advantage, the regimental hospital system
is the best. If this system had existed in the French army, they
never could have carried on their medical service, from the weakness
of their staff. If our surgeons were overworked, what must be said
of the French, who, with a much larger proportion of sick, had only
one medical officer in proportion to three of ours! Thus, at Constantinople during the winter of 1854-55, a French surgeon told me
he had 211 patients to see before 9 A.M., when, by their regulations,
the visit must be terminated.* With us, too, the continuance of the
war not only decreased the sick list, but augmented the medical
staff, so that at its termination, what between civil surgeons, assistant surgeons, acting assistant surgeons, and dressers, our hospitals,
especially those in the rear, were inundated by professionals of every
type, while the few who joined the army of our allies barely made
up for the vast mortality which their constant labor and exposure
to disease occasioned.

---

* The French were so ill off for medical attendants that they had to employ intelligent soldiers to dress not only simple wounds, but often stumps also, (under superintendence.) This most useful corps was called "Soldats panseurs."

by no means an agreeable winter's residence; but in summer, its airy position and the glorious view it commanded afforded a most agreeable and beautiful residence for the sick who were oppressed by fever and lingering convalescence on the burning and arid plateau. This hospital contained between 400 and 500 beds; and the results obtained in it were unequaled, in so far as curing disease was concerned.

Another *sanitarium* was formed above the monastery of St. George, at an elevation of 500 feet above the sea, and consisted of twelve large Chester double huts, each fitted up for twenty beds. The accommodation, both in the number of huts and in the number of beds which each could be made to afford, could be greatly multiplied on an **emergency.** The construction of the huts which formed this hospital was perfect, the ventilation everything which **was** desirable, the water supply sufficient, **the** kitchen arrangements most excellent; and, altogether, this establishment might have proved, if erected at an earlier period, one of the most useful, as it was one of the most **perfect** hospitals in the East. The beautiful scenery by which it was surrounded, the cool breezes which fanned it even in the heart of summer, the agreeable walks around, and the distance it was **from** the turmoil of the camp, combined to **render "the monastery"** as pleasing a residence as it was **a** favorable station for the sick.

The general hospital in camp, which might have been termed "the acute hospital," as to it the men struck down in the trenches were first carried, was well situated on ground elevated between 400 and 500 feet above the sea level, within the lines of the 3d Division, and close **to the** extreme left **of** our position. It consisted at first of **twenty,** and latterly of thirty Portsmouth huts. These huts were erected in four rows, facing west, leaving three broad streets between them. A space of about twelve to fifteen feet intervened between neighboring huts. This close packing was much to be regretted in the arrangement of the huts; but as they had originally been erected for the accommodation of the 14th and 39th Regiments, and as space was not easily procurable where so large an army had to be encamped, the error was in a great measure unavoidable. The ground on which these huts were erected had not been at first as carefully prepared or drained as it would have been if they had been originally erected for hospitals. Along the sides of each intervening street, deep ditches were dug, after its conversion into a hospital, to secure the drainage, and latterly the streets themselves were paved with round stones. A corps of Tartars was constantly employed in keeping the ground clean about the huts. The cookhouse and latrines were placed behind—the latter on the declivity of the hill leading down into the valley which bounded our camp from the higher plateau. These huts were erected during the winter of 1854–55, but they were not used for hospital purposes till late in the spring of the latter year. They were barely out of range, as some of the long shots from the well-known "crow's nest" battery came at times disagreeably near; but this propinquity favored the rapid admission of the wounded from the siege works. Each hut measured twenty-seven feet by fifteen inside, and contained, during the siege, fourteen beds; but, when the town fell, the number of beds was reduced to twelve. The air contained in these huts allowed about 260 cubic feet to each of fourteen patients. The total accommodation afforded by this hospital, during the siege, was 420 beds. The arrangement of the huts was as follows: At one end the door opened without the protection of a porch, (a grave fault;) at the end opposite to the door was a window; and, in some cases, there were also windows in the side walls, and a fixed one above the door. The beds were placed on either side, the heads being close to the wall, and the feet toward the center passage, which was three feet broad; one foot and a half of open space was left between the beds. There was a stove in the center, and ventilation-traps were cut in the sides, and in many cases in the roof. The openings in the sides could be closed at night, or in stormy weather. These huts being constructed of single boards, and roofed with felt, were not impervious either to rain or cold. In wet weather, water decks had to be constructed of waterproof sheeting. Many of them were completely floored with planks; but some had merely a raised dais on either side for the beds. Peat-charcoal or lime was frequently strewn beneath the planking, and the most scrupulous cleanliness was rigidly enforced. The men's kits were stowed in huts set apart for the purpose, so as to relieve the wards as much as possible from incumbrance. The bedsteads and bedding were excellent, the provision of medical comforts good, and the cooking passable; so that, on the whole, a better field hospital in the camp of an active army, I suppose never existed.

To a hospital so situated, one whose object was so temporary, and whose inmates were so liable to fluctuation, we cannot apply the same rules of criticism that we adopt when discussing the merits of more permanent establishments. That the patients were often crowded, that the proper amount of air was not measured out to each, that many of the refinements of a London hospital were wanting, may be admitted; but I doubt whether a better hospital could be provided 3000 miles from England, in a crowded camp, in a houseless region, before an active and energetic enemy, and almost within the vortex of the strife. The ventilation was much better, even at the times of greatest crowding, than could be supposed possible; and, as the wards were seldom full, except for short periods after some of the great battles or assaults, **it was** generally beyond cavil. Unquestionably the beds were too close; the huts were too near **one** another, and erected within the precincts of a crowded camp; and, in summer, the heat was great; but while I most willingly allow

5

that there was much in all this which was reprehensible, still I cannot conscientiously say that I had often reason to complain of the close air of the wards. The thinness of the walls of the huts, and the numerous air-traps cut in them, did much to prevent the formation of a dangerous atmosphere, but gave rise to disagreeable currents of air, of which the men often complained. This could not be effectively overcome by any lining, short of wood. The absence of porches to the doors greatly favored these draughts. In winter, the huts were very cold and uncomfortable, notwithstanding the pains we took to hang up blankets and bed-covers round the beds.

The routine followed in the distribution of the patients was as follows: The regimental hospitals received all those men of their respective corps who fell sick, and the wounded were also admitted into these erections during the early months of the siege, before the establishment of the general hospitals. Latterly, however, the large accommodation afforded by the general hospital, and its near neighborhood to the works, caused most of the severely wounded to be sent there, in place of being sent to their regiments. This remark does not apply to the light division, or the more distant parts of the right attack, except after the great explosion within their lines in November, 1856, when, from the destruction of the regimental hospitals, a large number of the injured were admitted into the camp hospital.*

During the ordinary course of the siege, assistant-surgeons, stationed with the troops in the works, paid the first attention to the wounded, before they were sent to the rear; but at the time of any assault, a staff-surgeon, in addition to assistants, was advanced into one of the ravines, and per-

---

* I have not spoken of the naval hospital, because it was not a military establishment; but, both in construction and management, it was one of the most perfect hospitals possible.

formed many necessary operations besides attending to the transmission of the wounded.*

For many months before the termination of the siege, the wounded were, with very few exceptions, operated upon in front, and kept in camp till nearly cured, when they were transferred to the large fixed hospitals on the Bosphorus and Dardanelles, or sent to Smyrna. If sick, a visit to one of the sanitaria was, latterly, often substituted for a voyage across the Black Sea. The ample accommodation and provision in camp enabled all this to be accomplished toward the end of the war; but, at the earlier period, both sick and wounded were sent at once to Scutari, where the required operations were in most cases performed. Thus, after the Alma, the wounded and those operated on were put on board ship within forty-eight hours, and so also after the 25th and 26th October, and especially after Inkerman, when the wounded were sent from camp at a very early date.

Thus, then, it is of much consequence to remember, in order to appreciate aright the surgical annals of the war, that, at two epochs of the siege, arrangements totally different were, from necessity, followed in the treatment and distribution of the sick and wounded. It can easily be understood what a difference such arrangements must have made in the mortality, after the receipt of such an injury as a compound fracture. How different it was to keep a patient so circumstanced, in a comfortable hospital in camp, and treat him among his comrades, from having him transferred, immediately after the accident, in a jolting carriage, over roads

---

* The French had ambulances in the ravines close down to the trenches, where operations were much performed, with dressing, the extraction of balls, etc. From these they passed on their wounded to the divisional ambulances, in one of which (the right) Baudens tells us they have had as many as 150 capital operations within twenty-four hours after one sortie, and in another (the left) they had as many as 400 wounded men carried in the course of one night, that of the 1st–2d of May, 1855.

full of mud-pits, and in the close hold of a crowded transport, over a sea proverbially stormy, to be disembarked, in a
manner the most faulty conceivable, at Scutari, admitted
among strangers into a great hospital at a period when the
fever occasioned by his wound was at its height, and placed
under the care of a surgeon who knew nothing of his previous history, or the particulars of his case!*   The deplorable effect occasioned by this early transference I will allude
to more particularly afterward.

The system of **nursing** which was pursued in our hospitals
at the beginning of the war was highly defective, and led, I
doubt not, to the sacrifice of many lives.   The attendants
on the sick consisted exclusively of soldiers, very often convalescents, whose strength was not sufficiently restored to
enable them to resume duty; and not unfrequently of the
worst set-up and most useless of the privates, whose presence
n the ranks could be easily dispensed with.   These were
detailed for the service of the hospitals in the proportion of
one attendant to ten sick, and in the transports, in the proportion of one attendant to twenty-five sick.   The necessity,
during the early part of the war, of having at his post every
man who could carry a musket, kept the number of orderlies
rather below than above this proportion, and often occasioned
the employment of men utterly unfit for the duty.   In hospitals at home, with few serious ailments among their inmates, the proportion of attendants allowed by the War-

---

* It was a most serious error in the transference arrangements
that no account of the patient was transmitted along with him, when
he passed from one hospital to another.   It prevented the surgeon
into whose care he fell from having the same interest in him he
otherwise might have had, and it obviously stood in the way of his
proper treatment, and destroyed most effectually all means of obtaining accurate statistics, or enabling one to follow cases to a termination.   The naval authorities managed this much better, in sending
a full account of each patient when he was transferred; but, of
course, with them this was more easily effected than it was with
soldiers.

Office regulations is ample; but in field hospitals during war the case is very different, as there the accidents are severe, and few of the patients are able to assist themselves. The distance at which the cook-houses and latrines are usually placed, in the arrangement of field hospitals, impose much labor on the orderlies, and the discomforts to which they are themselves exposed render them both mentally and physically less able to perform their duties.

The pensioners, who with so much parade of "unquenched courage" volunteered to serve in the ambulance corps, and tend the sick, proved most useless, being but little fit to undergo the fatigue which their duties entailed, and were, with few exceptions, sorely addicted to "a veteran's failing." They soon disappeared, being, to use a vulgar phrase, quickly **"used up."** I **saw enough** of civilian orderlies to impress upon me the conviction that they, too, **are** ill fitted for duty in military hospitals **where it is so** necessary **to** maintain both a military spirit and a strict military discipline. They do not understand the soldier, who, in his turn, despises them, and thus an antagonism, almost impossible **to** control, **springs** up between them. **Besides this,** the want **of the** previous training of military drill makes the pure**ly** civilian orderly by no means so manageable **or so "workable"** in a **camp where he feels as a** stranger, and withal, becomes **some**times a hopeless burden. The recently-formed medical staff corps of our army will, **I** believe, be found most efficient, being an imitation of the "Infirmièrs" of the French hospitals.*

---

* Steady soldiers of character, or men especially recruited, forming a distinct corps, having promotion granted them in it for merit, retaining the military spirit and the military idea, but still constituting a separate fraternity, well paid and *well fed,* (an essential for a sick attendant,) having **fixed duties** and regular training, not liable to military service, entirely under the command of the medical officers, and yet subject to military discipline,—such is a rough outine of the corps which would render the best service in the hospitals of an army in the field. Each hospital should have its own staff

Much has been said about the expediency of employing
female nurses in military hospitals, and though the question
admits of numerous arguments *pro* and *con*, on which I cannot here enter, still I have no hesitation in giving my entire
adhesion to the practicability and usefulness of this proposal,
if conducted with discretion and caution. I have seen much
of this experiment, and watched its working with attention.
I believe that Miss Nightingale has decided the question,
and that under her auspices an addition will be made to the
regularly recognized nursing arrangements of our military
establishment which will prove of the very highest value.
The great difficulty will **always be to find persons** fitted to

complete. Its non-commissioned officers (wardmasters) of various
grades, its cooks, and its tradesmen, and their distribution and emi-
gration should be entirely under the control of the principal medical officer, who should have it in his power to concentrate these men
or disperse them, as his ideas of the exigencies of the service dic-
tated, without reference to any higher authority further than the
responsibility which the due performance of the service entails.
The independence of the "Intendance" in the French service of the
medical staff, and, in fact, the superiority of the "Intendant" in
rank to the surgeon, and the authority which is given in their hospitals to the "Comptable" over the medical officers, is a decided
blot in their system. The medical man is thus deprived of his
proper position, and as his promotion depends on the Intendant, he
can hardly be expected to be bold enough to expose omissions in the
management. It is certainly very desirable that all the details in
the economy of a hospital should be completely managed by some
one else than the medical officer, whose special duties should alone
engage him, but the department (the purveyor's in our service) over-
looking these details should be entirely under the control and direction of the medical men. How to arrange the functions of the medical, commissariat, and purveying officers, so as to produce one
harmonious whole, and not have it, as at present, that each should
be called upon to serve half a dozen masters, is a problem not easy
to solve. The general hospital system is, of course, much more
favorable to the production of a good "whole," in this respect, than
subdivisions into regimental hospitals can be, and in this the advantages of such a system are recognizable.

undertake the duty. They must combine a vigorous body
with a well-balanced mind—a mind untinctured by vain "ro-
mance," but endowed with religious feelings of such depth
and strength as will enable them, "in the name of Je-us
Christ our Lord, in perfect charity and self-devotion,"* to
undertake their trying duties. There must be the most perfect subjection to superiors, no "fussiness" or nervousness of
disposition, calmness in the hour of danger, and, above all,
a large stock both of common sense and of cheerfulness. In
recounting these qualifications, I do little more than name
those so pre-eminently possessed by Miss Nightingale and
several of our leading nurses. It requires the complete ab-
negation of self-will which exists in the Romish church to
produce the *sœurs de charité*; but that the Protestant
church too can send forth a band equally efficient, the "deaconnesses" of the Rhine† and our own Eastern hospital
nurses have demonstrated. That most injudicious selections
were in various instances made for service in our hospitals,
few who had much to do with these establishments will deny;
but every unprejudiced and attentive observer must **acknowledge** the vast amount **of good** which the female nurses accomplished, and the incalculable service which they are capable
of performing when judiciously selected and properly organ-

---

* Instructions to **the** Superior of the Russian Institution for edu-
cating Sisters of Mercy.

† Those who are interested in this question will read with pleasure
the pastor Fleidner's personal account of the Kaiserswerth Institution for the training of deaconnesses which he has published, **as**
well as a short pamphlet which is believed to have been written **by**
Miss Nightingale, on the same subject, and published in London for
the benefit of the Invalid Gentlewoman's Establishment. **It shows**
that, long before the establishment of the order of **Sisters of** Mercy
by St. Vincent de Paul, the office of deaconness existed in the Christian church, as well as how energetic the **sisterhood** has been since
its revival. **The deaconnesses now supply a great** many hospitals in
Germany and elsewhere with nurses, and the system is rapidly spread-
ing to different countries.

ized. The respect, almost devotion, shown toward them by
the soldiers, disappointed those who prophesied different
conduct from "the British ruffian"

In military, more than in civil hospitals, is the kindly sympathy and gentle care of a woman required. The soldier
has no crowd of anxious friends or relatives to visit his sick
bed; but, oppressed by a sense of loneliness in a foreign land
—a feeling of which he is little conscious when in health,
but which comes upon him with overpowering weight when
the silence of the sick-room gives him time for reflection—
he clings to those gentle offices of kindness which a woman
can alone bestow. To the surgeon a good nurse is of incalculable service. His medicines and splints cannot cure
unless those many trivial and apparently insignificant points
connected with the management of a sick-bed, which are
more than half the cure, are attended to. The surgeon has
enough to occupy all his scant time in the greater and more
serious duties of his service; while those nameless, con-
stantly-recurring necessities of a sick-room, none can minister
to like a woman. To perform such duties aright seems part
of a woman's mission on earth.* The patience and gentle-
ness which are such inestimable qualifications in a nurse, we
cannot look for in a strong and vigorous man, nor yet that
sensitive recognition of a sick man's wants which so distin-
guishes a woman. Charles Lamb well expresses this idea
when he says: "It is not medicine, it is not broth and coarse
meats served up at stated hours with all the hard formality
of a prison; it is not the scanty dole of a bed to lie on,
which dying man requires from his species. Looks, atten-
tions, consolations, in a word, *sympathies* are what a man
most needs in this awful close of human sufferings. A kind
look, a smile, a drop of cold water to a parched lip,—for
these things a man shall bless you in death."

---

* Luther says: "A readiness to compassionate others is more
natural to woman than to man. Women who love godliness, generally have also a special gift for comforting others."

A woman's services in a hospital are invaluable, if they were of no further use than to attend to the cooking and the linen departments, to supply "extras" in the way of little comforts to the worst cases, to see that the medicines and wine ordered are administered at the appointed periods, and to prepare and provide suitable drinks. As to the employment of "ladies," I think they are altogether out of **place in military** hospitals, except as superintendents. As **heads** of departments, as organizers, as overlookers, "officers" of **the** female corps, if you will, they cannot be dispensed with ; but for inferior posts, strong, active, respectable, *paid* nurses, who have undergone a preliminary training in civil hospitals at home, should alone be employed.* In the camp

---

* Deputy Inspector-General Mouat, in his report on the Russian hospitals, (page 7,) says: "From what we saw and heard of these valuable women, (Sisters of Mercy,) with the previous knowledge of the attempt to introduce female nursing into our military establishments in **the** Crimea and at Scutari, we are led to **the conclusion,** irresistibly, that female nursing, as a general rule, **can be** only successfully practiced from either the predominance of strong feelings of devotion or affection. Founded on merely **mercenary** or any other feelings, it is not only liable, but nearly certain to fail if introduced into military hospitals; and such we believe will be the testi- **mony of most persons of any** experience who have carefully **attended to the subject** during the late campaign." It will be observed from **the text that,** claiming as I do no very limited experience of the working of the female-nurse system, I totally differ from my friend, Mr. Mouat, in his conclusions on one essential point, viz., whether the under nurses should be paid or not. Many arguments could be advanced to prove that they should be paid, and well paid too; but I will content myself by remarking, that I do not think we can get the class of women whose services are wanted in the **capacity under** review, to undertake these duties solely from **"strong** feelings of devotion or affection." Granting that the humbler classes are as accessible to such "feelings" as are they above **them in** rank, and that these prompt to many beautiful instances of self-sacrifice by their own "firesides," it is not to be expected that, with the circum-

hospitals, which, with an army in the field, are merely the temporary resting-places of the sick, men should alone be employed as nurses; but in the more fixed hospitals in the rear, the lady superintendents and under nurses should, in my opinion, always be added to the regular staff. Their attention should be limited to the bad cases, and they should have the entire control of the linen, medical comforts, and cooking. All cleaning should be done by men. There should be a lady superintendent over each division of the hospital, responsible to the surgeon as well as to her own lady chief. Then there should be a store of "extras" under her charge, distributable on requisition from the medical attendant, and which depot should be filled up to a certain quantity weekly, the *sister* being held accountable for the contents. Wine and all extras should pass through her hands. She should be responsible for the due performance, by her female subordinates, of their duties, and have a right to interfere with the wardmaster if the cleaning, etc. were

---

scribed views they have, they should feel themselves called on to go forth to strange and distant lands.

Even if we did secure the services of desirable persons whenever our armies go abroad, it is well known how difficult, nay, how impossible almost, it is to control them of that rank, when you have no other hold upon them than that founded on their feelings. With a lady, many other influences come into play. It becomes absolutely necessary to make the appointment "a good situation"—one to be coveted and not lightly lost—in order to secure the services of, and exercise the proper control over, the class of individuals wanted to fill the inferior posts My impression is, that, *as a body*, the paid nurses proved better during the late war than the unpaid; and I suspect a much higher authority than mine could be found for the statement. I do not for a moment wish to assert that religious character is not an almost essential qualification to be sought for in the nurses of all grades; but I do not think that we can make the system work, if we rely on religious feeling alone as the moving agent to the arduous services required. All the under nurses should, in my opinion, obtain a handsome stipend.

not properly attended to by his male corps. Both wardmaster and sister should be accountable to the surgeon of the
division.*

There is no part of the organization of a hospital which
demands more attention than the **cooking department** and
the proper distribution of the food, and none which, in military hospitals, is, in general, more neglected. The truth of
the maxim which says, " La première condition de la santé
c'est la satisfaction de l'estomac," is beyond question. Those
who remember the cooking for the sick which prevailed at
Scutari before, and that introduced after the kitchen department underwent the "female revolution," will be able to
appreciate the difference which attention to this point must
make on the results of treatment. A morsel which disgusted
the healthy could hardly have been relished by the invalid,
and when one calls to recollection the " portions" which
were dealt out to the sick in those days of cloud, he cannot
wonder at the awful mortality which reigned.† Men turned
away with loathing from the coarse, half-cooked, **tough**

---

* In the cleaning of the wards, and attending to the sick, I found
it a good plan, when organizing my division of the Smyrna hospital,
to divide the beds into batches of sixteen, to each of which two
male orderlies were attached. Each of these men was made responsible for eight beds, with the floor and utensils included in the space
occupied by that number of beds. One of these men was called the
"diet orderly," as his duty included the bringing to the ward, and
the distributing of the rations to the whole of the sixteen patients
included in his division, and he also attended the sister when she
distributed the wine and extras. The other man was the "medicine
orderly," who had charge of the dressings, and went for and distributed the medicines ordered at the visit. I found eight beds as
many as one man could properly arrange, and I confined the service
of the two female nurses I had in my division to attendance on the
severe cases, the preparation of suitable drinks, the administration
of medicines and extras, and the changing of linen. **The** duties of
the sister I have sketched in the text.

† At the time to which I refer, the deaths frequently ranged above
sixty a day at the Scutari hospitals.

morsel, which, even if consumed, was incapable of providing
them with the nourishment they so urgently required. It
was in the management of those cases of such frequent occurrence in the East, where a lingering convalescence—most
liable to relapse—had succeeded to a wasting flux or debil-
itating fever, that the "extras" from the "sisters' kitchen"
came to tell in the treatment. Nourishment, properly and
judiciously administered, was the sole medication on which
we could rely in such cases. It was often of itself sufficient
to cure, and it was in attending to this that the female nurses
saved so many lives. I shall not soon forget the change
effected by the offices of the female nurses in my division at
Smyrna, by the careful regulation of the diet alone. I do
not hesitate to say that, previous to their arrival, I had lost
many patients whose lives I might have been the means of
saving if I had had then such assistants. Though much remains to be said on this subject, my space forbids entering
upon it further.

The **transport** of the sick and wounded connected with
an active army **is** always **a** matter of difficulty, and **is not**
uncommonly the indirect source of increasing the mortality.
For the army of the East, the provision made by the medical
department was most ample ; but, unhappily, military " sa-
gesse" did not always recognize the importance of its being
carried out. Hence, after the Alma, the medical staff had
helplessly to deplore the fatal abandonment of the forty ambulance **wagons provided for the** service. In consequence
of this measure—strongly remonstrated against by the medical staff—the sick and wounded—those suffering from cholera, or from broken bones and amputated limbs—had to be
carried some miles to the beach, under a scorching sun, either
in blankets slung between two oars, or jolted over tracks
deep with sand, in the most uncomfortable of all earthly
conveyances, Turkish arabas. The poor fellows were then
crowded into the hold of the transports, or laid in rows on
the hard **deck,** with scarcely a single attendant to answer

their piercing cries for water, or for a blanket to cover them.
That passage to Scutari is as one of the wildest nightmares
which ever disturbed an excited brain. Numbers sank on
the passage, and many died afterward from its effects.*
After the establishment of the siege, hospital wagons and
"cacolets," or mule litters, conveyed patients to Balaklava,
where they were shipped for the Bosphorus or England.

When the land transport corps was fully organized, the
road to the front finished, and the splendid line of steamers
established between the Crimea and the Bosphorus, a transport service existed such as, I suppose, was never before seen
in the history of war; but before this perfection was attained,
much suffering had been undergone, and many deaths had
been occasioned. The scant conveyance, rutty roads, and
foul ships of the first period are now, with many of the
earlier miseries of the war, almost forgotten and unknown.
Even at the best of times, all who have watched the effects
of the transport will, I think, acknowledge the malignant
influence it directly and subsequently exerted on the wounded. The jolting in passing from camp to port, the hoisting
in and out of ship, the close air of the hold, the irregular
feeding, the sea-sickness, the decreased attention to dressing,
and the thousand **and** one hardships to a sick man of **a**
conveyance by land and sea,—all combined to influence the
wounded most prejudicially.† It was well if the patient es-

* Out of 1300 embarked at Old Fort, 51 died during the short
passage to the Bosphorus. The ambulances were so constructed as
to carry some men sitting, and others on stretchers. The mule
litters also could be made to convey patients in either position.

† This is strongly confirmed by Dr. Jenkins, in his report on the
naval brigade, when he says: "We ought not to forget the necessity
there existed for removing the men to Cossack Bay within a day or
two after operation, and the effects of such removal. Those who
left the camp in a favorable condition have arrived at Balaklava in a
state of delirium; and stumps which looked well in the camp have
been found to be in a state of inflammation when the patients

caped the deadly typhus which lurked in these ships, and
which brought death to many who had embarked in comparative health.   If he had a broken limb or a suppurating
wound, it was more than likely that an unhealthy condition
would be engendered by the accidents of the passage; and
it was lucky if gangrene did not ensue, and amputation follow.   The French ascribed much of the gangrene which
reigned in their hospitals at Constantinople to the early
transference of their wounded from the camp—a measure
which was adopted by them in order to maintain the *morale*
of the soldiers.

Not a few of our men sank under the trial of the voyage,
and lie buried beneath the restless waves of the Euxine;
while to stand, as I have often done, on the pier at Scutari,
and watch the landing of the survivors, gave one a most
vivid idea of the pernicious effects which the voyage occasioned.   Many I have seen die while being landed; and I
remember six men in one morning being disembarked in life
at the pier, and dying before reaching the hospital.   The
complete prostration and exhaustion written on the faces of
all told in characters which no words can express the severity of the suffering through which they had passed.

reached their destination. In short, the consequence to a fresh-
formed stump, of a three hours' jolting over a bad road, even in the
best slung ambulance, may be easily imagined. The evil effects of
the journey to Balaklava upon men who were not considered fit cases
for operation were so obvious that latterly, excepting for the slightly
wounded, ambulances were never used, and all the men who had
undergone severe operations, or had been badly wounded, were conveyed on stretchers borne on men's shoulders." **Dr. Davidson, of**
Therapia hospital, says, in his part of the same report: "So utterly
prostrated were these men (patients) when they arrived, **that the
wonder was, not that so many died, but that so many recovered."**—
Report on Baltic and Black Sea Fleets, pp. 37 and 47.

CHAPTER III.

The Campaign in Bulgaria, and its Effects on the subsequent Health of
the Troops—The Diseases which appeared there, and during the Flank
March, as well as afterward in the Camp before Sebastopol.

IN the previous chapters I have described the position
occupied by the British army before Sebastopol, as well as
its condition in regard to food, clothing, and duty. To
avoid confusion, I have hitherto omitted to trace the incidents of the earlier part of the war, as they bore on the
health of the troops; and I have not yet mentioned the
diseases which prevailed at different seasons in our camp.
Before entering on the subject proper to this volume, I must,
however, advert to these points, though very briefly, as
their effects on the constitutions of the men, and on their
recovery from wounds and injuries, were both great and long
continued.

It is easy to infer that the adverse circumstances before
mentioned must have had a powerful influence in giving rise
to many of the diseases which appeared, as well as in causing the strange rebelliousness which marked some, and the
absolute incurability which characterized others.

Early in June, 1854, as gallant and splendid an army **as
Britain ever equipped landed at Varna.** Its numbers,
amounting at that time to 15,000, were soon augmented by
the addition of upwards of 10,000. For **nearly** three
months it acted as an army of observation, having its encampments scattered between the coast and Shumla, and
lying in close neighborhood to large bodies of French and
Turks, whose combined force could **not** be less than 30,000

men. This large body of troops was chiefly concentrated
about Varna, stretched along the valley behind the town, or
crowning the neighboring heights with their white tents.

The country in which our army lay was of the most beautiful description to the eye, but of the most dangerous character to the health. The long, shallow lakes, exuberant,
low-lying woods, thick-tangled wild flowers, and verdant
grass presented to the eye of all but the initiated one of
the most charming stations for an inactive army. But to
those who judged of encampments by other tests than those
of natural beauty, all these charms were but as the fair
painting on the cheek of death. Every element of its physical character was bad. Large surfaces of shallow water,
surrounded by level, spongy lands, indented with little hollows, dried and cracked by the recession and evaporation of
the winter floods; low brushwood, rank in vegetation;
bounding uplands,—these, with a high temperature, and a
deficiency of potable water, supplied nearly all the possible
combinations of physical destructiveness.*   When these
features failed to warn, the ominous designation of "the
valley of death," which the natives gave to part of the
locality, could not prevent the "experimentum crucis" being
made, and so our splendid army was placed within the vortex of these baneful agencies, and its fate was not long left
in doubt. The crowded burying-place of the Russians in
the "old war" formed part of the encampment occupied by
the British, and thus friend and foe lie mouldering in the
same graves.

During the period of the year when our army occupied

---

* "The experience of all ages has proved that the neighborhood
of marshes, grounds subject to be overflowed by large rivers, surrounded by foul, stagnating water, or low places covered with wood,
are most injurious to health, and the noxious effluvia arising from
these situations are augmented in proportion to the heat of the
climate or the season of the year."—*Sir George Ballingall's Outlines*,
p. 50.

Bulgaria, the two most prominent climatic features were a hot sun in the daytime, (90° to 98°,) and cold, dewy nights. The heavy mists which rose from the steaming lakes in the valley spread their heavy mantle over the camp at night, and introduced into the bodies of the unconscious sleepers the seeds of future disease and death. That many who escaped the immediate effects of such a residence, then imbibed a poison which afterward showed itself in their behavior under injury, few will question, who watched the phases of disease during the subsequent periods of the war.

The rationing of the troops, too, when in Bulgaria, was bad and irregular, the tents thin and permeable, and an ample supply of deleterious spirit and adulterated country wine at hand. These all lent their aid in predisposing to the outbreak of disease.

The apathy which these causes, and, above all, the want of employment, engendered, all tended in the same direction, so that when cholera broke out in July, a better field for its ravages could hardly be imagined. The power which terror has in propagating this disease received many most striking exemplifications at this time. The French and Turks suffered most. The horrors of their hospitals recalled the pictures of Boccaccio. Half of the **army** of Espinasse, in the Dobrutscha, disappeared as by a whirlwind, and the panic which seized the survivors has been described to me as having been beyond belief. This cholera was the great scourge which devastated our camp; but typhus, its close companion, diarrhœa, and dysentery, all claimed their tithe of men. The breaking-up of the large encampments failed to rid us of the enemy, which clung to our army with a pertinacity and malignancy that nothing could overcome. Thus, then, in less than three months we find that 897 **died** from cholera, and 75 from dysentery and diarrhœa. The Light Division, the Guards, some of the heavy dragoon regiments, and the commissariat department appear to have been the heaviest **losers.**

6*

It is not, however, for the purpose of repeating the tale of the heavy losses of our army in Bulgaria, that I make these remarks. It is in order to indicate what a weakening and deleterious effect the residence in that country exerted on the survivors, and how much its effects must have told on the issue of disease and accident afterward. There is no one fact which more completely illustrates this pernicious influence, than what all surgeons who served in Bulgaria will remember, that numbers of men, without being absolutely diseased, or yet so ill as to be fit for hospital, or perhaps even for medical treatment of any kind, yet fell off in appearance, lost appetite, flesh, and color, became listless and weak; and that almost every one who had seen the campaign out, was conscious of a considerable difference in his state of health after he landed.* The standard of health, in short, was lowered, the vital forces were diminished previous to embarkation for the Crimea in September; and though a state of sickness had not been established yet, that prelude to it existed which wanted only a determining cause to develop it.

This "unsatisfactory condition" was well shown during the short marches to the place of embarkation, by the num-

---

* The very serious effect which inaction has in determining disease in an army needs no illustration. The annals of the Peninsula afford many examples. That a much less proportion of sick existed in the Peninsula when the army was fighting and marching daily, than existed in our army in Bulgaria when no duty almost had to be performed, is only in keeping with many other facts of the same character spread broadcast over the pages of history. "My estimates lead," says Dr. Aitken, "with still greater force to the conclusion, that the amount of sickness at Varna was greater than that of the French army in Spain, and nearly as great as the army of Portugal while engaged in very active campaigns, and this, too, though not a soldier in Lord Raglan's army had fired a shot." "The period of smallest loss to an army," says Mr. Alcock, "is a victorious and vigorously prosecuted campaign, with frequent battles and much marching."

ber of men who "fell out," and the large proportion who were unable to carry their packs. It was also officially recognized in the order by which the commanding officers of regiments had the option of making the men land without their kits, on disembarkation at Old Fort.

There is no doubt that the spirit infused into the men by the prospect of employment had a good effect on their health at the period of embarkation for the Crimea, but still the cholera did not leave them. From some of the transports burials took place daily during the transit to Old Fort. The heavy rain which drenched the unsheltered army during its first bivouac gave the disease a fresh impulse; and when, after the Alma, one of the first principles which regulate encampments was violated by our men being halted on ground lately occupied by the enemy — an enemy, too, among whom cholera had prevailed — and camped amid the dirty straw, old rags, and filth which the Russians always leave behind them in such profusion, the disease broke out with violence.

In the vineyards of the Balbec our soldiers ate voraciously of the grapes which there hung in such tempting clusters, and drank immoderately of the streams which, splashed and muddied by the hot wheels of tumbrils and guns, and the dusty feet of many men, filled their parched mouths with sand. Connect all this with the most exhausting fatigue— a fatigue which caused the immediate death of some—with the dreadful heat, the excitement, the want of food and sleep, and then it will be easily understood why 596 **men** sank during the famous flank march, and 2237 were sent off sick therefrom to Scutari,* as well as that many arrived before the city utterly exhausted, and that many never fairly got over the effects, engaged as they were almost immediately in their arduous trench duties, and thus deprived of any

* See paper by Dr. William Aitken in *Glas. Med. Journal* for April, 1857.

opportunity of repose or recovery. Those men whom ill-
ness or fatigue prevented from keeping up with their comrades were left behind to take their chance, as there was no
conveyance to carry them on; and both from this cause, and
from the deficiency of opium in the chests to meet the demand, many lives were lost.

Thus, then, in the short period of three months in Bulgaria, and the twenty-two days which elapsed between the
landing in the Crimea and the camp being formed before
Sebastopol, a very large number of lives were lost, and the
seeds of much of that sickness sown which yielded such a
rank harvest afterward.

The troops were without tents for some weeks after landing, their packs were not returned to them for nearly two
months after the establishment of the siege, and their squad
bags not having been forwarded from Scutari, "the soldier
was left during the interval almost in rags, a prey to vermin,
and without a change of any kind," (Tulloch,) all this while
undergoing great fatigue and much exposure.

Thus, then, I have rapidly sketched the progress of the
expedition up to the sitting down of the army before Sebastopol, and indicated the leading circumstances which exercised an influence on the health of the troops. Let me now
shortly inquire what were the diseases which resulted from
these circumstances, and from those other conditions that
came into play at a later period—already traced in a previous chapter—which all combined to destroy 10,000 men of
our army in seven months, and to delay the fall of the fortress for a year.

It is a remark of all times, that disease thins the ranks of
an army far more than is done by the arms of an enemy.
The ignorant and unreflecting dwell more upon those events
which are of unusual occurrence, while they pay but little
attention to the effect of those causes with whose action
they are familiar. Thus it is, that to the unthinking observer, the ravages of battle present themselves with greater

force than the more obscure but more deadly influences of disease.

The proportion in which the victims of disease will exceed those of battle varies with the country and the season in which the campaign is waged, as well as with the resources of the army engaged. It was a saying of Frederick the Great, that fever alone cost him more men than seven pitched battles, and it has been an axiom with most commanders, that "more campaigns are decided by sickness than by the sword." At Walcheren, in 1809, our army, numbering 40,000, lost 332 in the thousand by disease, and only 16·7 by wounds. In the Peninsula, from January, 1811, to May, 1814, during which period the battles of Albuera —"the most desperate and bloody of the whole revolutionary war"—Salamanca, Vittoria, the Pyrenees, Nivelle, Nive, Orthes, and Toulouse were fought; and Badajos, Ciudad Rodrigo, and San Sebastian stormed, besides many lesser encounters, in an effective force of 61,500 men, only 42·4 per 1000 were lost by wounds, while 118·6 were lost by disease. In Burmah, again, under a less propitious climate, in the first war 35 per 1000 were lost by wounds, and 450 per 1000 by disease; and in the following year 106·6 per 1000 were lost by battle, and 300 per 1000 by disease.* In the Indian campaigns, in the wars of the Empire, and in the Russian campaign against Turkey in 1828, the difference is still more marked. To multiply examples would be of little use, but sufficiently easy, as the same is the teaching of almost all campaigns.

In the Crimea the proportion of those lost by sickness to those lost by wounds was, if we take the whole war, as 16,211 to 1761;† and if we calculate merely the **period**

---

* Bauden's Hygiène Militaire comparée. The experience of the French in Egypt is an exception. There they lost 4157 by disease, and 4758 by wounds and accidents, in an army of 30,000, and in a period of three years and a half.

† Not including those killed in action.

from October, 1854, to April, 1855, then, in a mean strength of 23,775 men, we have 9248 lost by sickness to place against 608 lost by wounds;* or if we extend the period, and number the loss in strength by disease and wounds from October, 1854, up till the conclusion of peace in 1856, then the contrast becomes yet more marked and decided. In the French and Russian armies engaged during the war, I believe the proportion of disease to wounds to have been even higher than in ours; but the want of accurate details prevents any close approximation to their loss being made. M. Scrive tells us that in December and January, 1854–55, their admissions into the Crimean ambulances were 15,500, of whom 14,000 were for disease, and 1500 wounded; that of the whole number 1700 died; and that during the last six months, in which the final assaults that led to the taking of the city were made, the French had 21,957 wounded, and 101,128 cases of disease.

It has been calculated that during a campaign, an average of 10 per cent. sick may be looked for; but in the Crimea the average was very much above this, as at one time, (October, 1854, to April, 1855,) even though our army was stationary, in a comparatively healthy climate, having its communications open, and within a few miles of the sea, the percentage of sick to strength rose to 39 per cent. for the whole body of infantry, to 45 per cent. for the troops serving in front, and for eight corps to the unheard-of number of 73 per cent The average percentage of sick to strength during the whole war cannot be ascertained with precision.

The diseases which chiefly affected our troops were cholera, diarrhœa, dysentery, typhus, and typhoid fevers. It is true that some of these were often so mingled, so confounded in their manifestations, and so modified by their mutual reactions, that it was not always easy, nor sometimes possible, to detect their individual influences, or their mutual correla-

---

* Tulloch's Crimean Com., p. 152.

tions and interdependence on one another; yet in many cases the distinctions were well marked at first, or their invividuality was shown during treatment. The scurvy poison was the fusing medium, if I may so express myself, which blended the one disease into the other, and modified all; and as the affection was more developed among the French than among us, I believe the confusion of nosological diseases to which I have referred was more striking with them than even with us.* This curious confounding of affections often put one's previous notions of disease completely at fault, rendering the diagnosis uncertain, the indications for treatment curiously at variance, and combining the pathological results in unaccustomed synthesis.

The intermediary position which the Crimea occupied between Europe and Asia seemed to have caused that correlation of disease from which our troops suffered. The typhoid fevers of European armies, and the violent dysenteries of Asian and African, there met and struggled for the ascendency; while the omnipresent cholera ravaged our ranks, and scurvy, the product of no particular clime, but of man's own improvidence, prepared the way for their several assaults.

In the returns of the war, 4513 deaths appear from cholera. The whole number treated was 7575, giving a percentage of 59·57 deaths on the whole number.† The great majority of these occurred before leaving Varna; but subsequently to that period two distinct outbreaks of the epidemic took place—one in December, 1854, and the other in May, 1855. This was the disease which chiefly attacked the new drafts, among whom, however, it was not so deadly,

* My friend Professor Tholozan has given a very able exposition of this mixing in his paper read to the Academy of Medicine in September, 1856.

† These numbers do not include officers, of whom 147 died during the war of sickness, and 86 of wounds.

in proportion to those seized, as it was among the old cam-
paigners. The period when the disease was at its greatest
height in the Crimea was in December, 1854, when 888
cases, in an average strength of 29,727, appeared, and 636
died—giving 2·9 per cent. cases of the whole force, and 71·6
per cent. deaths on admissions. In the June epidemic the
French had 5450 cases of cholera, of which 2730 died. Its
rebelliousness to treatment was always marked during the
period of intensity, and its manageableness during its retro-
cession. There seems good reason to suppose that the
cholera which appeared at Varna was introduced from Mar-
seilles, where it was then prevalent; and while for considerable periods it remained quiescent, it never wholly ceased in
the camp, until its disappearance in the end of February.

Scurvy was the great destructive agent against which it
was most difficult to cope, and which, though but little cog-
nizable by its usual signs—though often carefully masking
its presence behind some other ailment—yet influenced every
disease, and touched with its poisoned finger every wound.
Sometimes breaking out with a malignancy which recalled
the graphic descriptions of the early voyagers, or the mas-
terly delineations of Lind, but more commonly declaring its
presence by a more negative, though not less baneful influence, in preventing or retarding cure. Wasting fluxes,
occurring during the treatment of an injury and defying
cure; hemorrhages of frequent repetition and difficult suppression; fractures refusing to unite; sores unaccountably
slow to heal, were its most ordinary indications to the surgeon. With us it was not often the immediate cause of
death; but with our allies it very frequently was. During
the cold weather they lost many scorbutic patients rapidly,
from effusion into the lungs, and suffered far more from
scurvy than we did. The sloughing buboes in the axilla and
groin which they had to contend with were among its worst
complications. M. Scrive says that its outbreak was most
marked **in** the regiments newly landed, in the proportion of

25 per cent. to 10 per cent. among the older regiments, and that it was less in the besieging than in the observing army in the valleys in the rear. With us, the earlier and most usual symptoms of the scurvy were a weariness of body which indisposed to exertion; a feeling of despondency; some degree of dyspnœa; stiffness in the limbs increased by rest, and relieved, in a great measure, by exercise; hardness of the muscles of the calves of the legs, the integuments of which were discolored so that they looked as if peppered with gunpowder; and at times, puffiness of the extremities. Such symptoms were often present, when the bleeding gums and other more serious and decided indications of the poison were absent.

The omission, before adverted to, with regard to the distribution of lime-juice, had a great influence on the development and progress of this dyscrasial disease. Even with salt food, if a sufficiency of anti-scorbutic remedies had been provided, such as well preserved vegetables, in default of potatoes, or other fresh legumes, lime-juice, or sour-crout, or, if the large cabbage of Turkey, greatly praised by the natives, had been freely distributed, the ravages of this affection might have been prevented or stayed. Who can calculate the number of lives which were sacrificed indirectly, if not directly, through such omissions! for this scurvy was our worst enemy, and, in truth, wrested from us more wounded men than even the conical ball.* The French used the indigenous dandelion largely, and if the perseverance of their soldiers in seeking it in every recess of the plateau was recompensed by its effects, they must have benefited largely

* Since such fruit as apples could have been procured for the troops in large quantities, it is a pity the authorities **were** unmindful of the experience of Virgil, who tells us, in the 2d **Georgic,** v. 130, of the "Felicis mali":—

 * * *   "quo non præsentius ullum
 * * * * * * *
Auxilium venit; ac membris agit atra venena."

by its use. Iron appeared to me to have more power over
this disease than any of our other remedies. Under its use
the blood assumed a more normal condition, and the health
of the patient greatly improved. The French put great
faith in the external as well as internal use of lemons. The
influence of scurvy in causing that curious eye affection,
hemeralopia, was frequently evinced. It appeared at an
early date among the Sardinians, and was not uncommon
with us. Its connection with deficient nutrition, or with
a degenerated condition of the blood, was thus rendered
apparent.

Of diarrhœa and dysentery, 52,442 cases were admitted
into hospital during the war, and 5910, or 11·26 per cent.,
died; while of these, 23,149 cases, and 1999 deaths occurred
in the Crimea before April, 1855. The presence of scurvy,
the use of irritating food, together with the labor and exposure, sufficiently account for this high mortality. Dr.
Tholozan tells us that 700 or 800, out of a total of 1200
cases which fell under his observation in the Pera hospital,
during the winter of 1854–55, had suffered from diarrhœa
or dysentery at the outset of their several ailments. In a
third of these cases blood had been passed, and in seventy-
nine autopsies the large intestines were engaged sixty-three
times, the small forty-two times, and the stomach thirty-
eight times. From April, 1855, to June, 1856, comparatively few cases of dysentery appeared, showing how greatly
the improved hygienic condition of the army influenced the
development of this "camp pest."

Very few, indeed, who served in the Crimea throughout
the first winter escaped an attack of dysentery; and it is in
keeping with my observation, that most of those who escaped
entirely were officers who seldom **ate the** salt pork, but who
subsisted on fresh food, which their private means enabled
them to procure. The proportion of officers to men who
suffered from either of the complaints specified, up to April,
1855, I have not been able to ascertain.

The prevalence of **ulceration of the intestines**, especially toward the lower part, was perhaps the most constant of all the pathological conditions found on post-mortem examinations in the East, and was almost universal in cases of enteric disease. The immense majority of those who served during the early part of the war were so affected, the ulceration being rather of recent than ancient date; and this remark does not apply to those alone who died of abdominal affections, but also to those who succumbed from other diseases, or from wounds. It is also a fact, which I have had many opportunities of verifying, that men killed in action at a time when they were apparently in the possession of health, or rather, as it should be put, men dying shortly after receiving severe wounds when seemingly robust, were found to have ulcers in their intestines, sometimes of a very extensive character. To this it was not uncommon to find diseased kidneys and lungs added. The disease, in these cases, might not be active at the period of death, but it was ready to break out whenever any injury or operation made an extra demand on the powers of life. It is of importance to note this extraordinary prevalence of undeveloped disease—this deceptive character in the appearance of the men—as bearing on their behavior under accident. The examples afforded by post-mortem examinations of intestinal ulcers in all stages of increase and of cure were many and interesting.

The influence which the intestinal flux had, when combined with scurvy, to modify and restrain the other manifestations of the blood disease, and the marked manner in which abdominal affections appeared to prevent the development of thoracic disease of a tubercular description, were well exemplified in the Crimea. This derivation, as it may be termed, had more to do with the striking immunity from phthisis, which prevailed, than any goodness in the climate, as, if it had not been for this counteracting agency, the other exciting causes, to which the troops were exposed in such abundance, would have been more than sufficient to over-

balance the advantages of any climate. Of phthisis, only
279 cases appear in the returns during the whole war, and
116 deaths in the East. I know, however, that many who
there showed no symptoms of the disease, subsequently
succumbed to it.

Of fever, (not typhus,) 30,376 cases and 3161 deaths
therefrom appear in the returns; March, 1855, was the
month when it most prevailed, and 21·0 per cent. the mortality during that month, and 10·4 the percentage of mortality in the cases treated throughout the war. In this
"Crimean fever," there was nothing whatever peculiar, unless the absence of any great febrile action, the rapid prostration, slow convalescence, and proclivity to relapse be
taken as specialties, which they were not, in my opinion, as
in nothing did this fever differ from the typhoid fever seen
in large cities, especially in Paris. It often followed dysentery, which, by reducing the patient's strength, prepared the
way for this "fever," as it was termed, but which, in these
cases, was the mere development of great vital prostration,
with the complications that were to be looked for in such a
sequence. The characteristic spots were not always present,
relapses were frequent, and during convalescence, tubercu-
losis was not uncommon. There was always a strong tendency to this fever evinced in our army. The ease with which
it was engrafted on other ailments indicated the fall in "the
health barometer."

Typhus fever killed 285 out of 828 cases. The spring
months of 1855 was the period when it was most prevalent.
It was the true maculated typhus as seen at home, with its
measly eruption appearing on the seventh or eighth day, and
not unfrequently complicated with pneumonia. This disease
was by no means common in our army, but its ravages were
dreadful among the French and Russians. The much greater
crowding which existed in their hospitals probably accounts,
much as any other cause, for the difference. The French
died of it by thousands, and the Russians by tens of thou-

sands. Neither I, nor, I believe, any other person, can tell exactly how many thus perished, and there is little use in speculating on numbers. The scurvy played here a most important part, as, when it was much developed, the fever was incurable. I cannot say that I had occasion to notice those marked remissions in the fever of the camp, of which several medical men have spoken; but I am persuaded that the treatment by large doses of quinine seemed to have a manifest effect over the low fevers of the early part of the war.

With intermittent fever we had little to do; but the French, who were stationed along the Tchernaya, suffered greatly from it, as did the Russians in the valleys on the opposite side of the river. I am not aware of anything whatever peculiar in this fever as it appeared among them; but I have heard from their surgeons that men subject to it were most unpromising patients, if injured by gunshot, especially if they combined any of the scorbutic taint with the paludal poison.

In autumn, jaundice was very prevalent, though commonly slight and easily curable by a visit to the Bosphorus; still it was sometimes very severe and intractable. The mere change from camp life and feeding, which took place when sent to sea, or to Scutari, commonly "did the doctor" sufficiently.

Perhaps the symptom which most struck the casual visitor to the hospitals during the winter of 1854–55, was the *anemic appearance* of the men. Their blood had been so completely depurated, that they had often more the appearance of chlorotic females than of soldiers. It was impossible, in a great measure, too, to get the defect supplied. No treatment almost effected any change, and thus it came to be a most serious affair if any hemorrhage or suppuration had to be encountered.

In this rapid review of the diseases of the camp, I have had no wish to be in any way minute. All I intend by refer-

7*

ence to them, is merely to indicate those which prevailed,
that so their bearing on the surgery of the war might be
appreciated. It is clear that many of them owed their existence and fatality to vicious hygienic conditions, of whose
influence, in deteriorating the constitutions of the men, the
presence and progress of these diseases afforded the best
proof.

That these diseases depended on the unfavorable circumstances as to food, shelter, and duty in which the troops
were placed, has been clearly demonstrated by Sir A. Tulloch, as he shows that the mortality varied in different corps
in an exact ratio with the care that was taken of each in
providing them with good food and shelter, even though
performing severe duties; and that those troops who were
constantly in the trenches, and badly supplied with clothing
and food, suffered most. Thus, while the mortality among
eight corps in front was as high as 73 per cent., and among
the infantry generally employed in the trenches 45 per cent.,
yet, in the naval brigade, who were always engaged, but
were well housed, clothed, and fed, it was under 4 per cent.
Among the cavalry, who, though perhaps not over-well fed,
had yet no trench or night duty, it was 15 per cent., and
among the artillery, who were well looked after, and less
worked, it was 18 per cent.; while among the officers, who,
though equally exposed, had yet the means of obtaining
better food and clothing, it was only 6 per cent.

The trench duties had certainly most to do with the
mortality, as its dependence on the length of time during
which these duties had to be performed was very marked.
I will not repeat here details with which all are familiar.
Sir A. Tulloch has entered into them at **length. In**
January, 1855, the sickness had attained its maximum; at
that period the number of those in hospital and at Scutaria
exceeded the force fit for duty, being as 12,025 **sick to**
11,367 effective.

I was always strongly convinced that the Bulgarian

campaign exercised a great influence, not so much on the proclivity to disease as on its fatality when formed. This has been most clearly shown by Dr. William Aitken, in his interesting papers on the health of the troops at the period implied. That the effect of injury and operation on these men showed how much their constitution had suffered during their residence in Bulgaria, I had often reason to observe, and an examination of the returns shows that "while the admissions to hospital were so much greater among the Crimean portion of the army, the deaths per cent. on these admissions were very much greater among the ex-Bulgarian part;" and that while the Crimean portion suffered chiefly from enteric and scorbutic disease and cholera, the Bulgarian troops suffered from fevers and pulmonary diseases, which is just what might have been *a priori* expected. The mortality on admissions from fever was nearly double—from cholera, dysentery, scurvy, frost-bite pulmonary disease, much higher—among those troops who served in Bulgaria than among those who were only in the Crimea. I cannot say how far the paludal poison of the swamps of Varna had to do in predisposing the troops who had imbibed it to fevers of a typhoid type. Dr. Aitken appears to give considerable weight to such a predisposition, and quotes the results of the Walcheren expedition as affording an analogous instance. There occurred several well-marked instances to prove that many soldiers, even some who never had any of the symptoms of miasmatic poisoning when in Bulgaria, showed signs of such an invasion when reduced by wounds; and I had frequent occasion to remark, that the advent of purulent contamination bore in such men a more than usual resemblance to an attack of **marsh fever.** That the **subtle influ**ence of this poison had been absorbed, and afterward prejudicially affected many who did **not at first** show signs of its presence, cannot be doubted by those who had much opportunity of observing the progress of disease and of wounds in the hospitals of the Eastern army.

There was an affection of the hands and feet very common
during the first winter and spring in both the English and
French hospitals, and which Dr. Tholozau, with some of the
French surgeons, was inclined to look upon as allied to a
peculiar disease that appeared epidemically in France between the years 1828 and 1832, and then termed acrodynia.
I feel persuaded that with us it was the product of cold and
scurvy, and was perhaps a junction of rheumatism, or yet
more probably of an early stage of frost-bite, with a weak
circulation and a scorbutic taint. This affection showed
itself chiefly in the pulpy parts of the feet and hands, but
especially in the ball of the toes, in the edges of the feet,
and in the muscular ridge which runs across the sole of the
foot at the roots of the toes. Its earliest symptom was a
prickly sensation experienced when the patient stood on the
foot, and was variously described by them as resembling the
pricking of pins, or as if they walked on nails. There were
lancinating pains in the calves of the legs, which parts felt
hard and brawny, and were sometimes swollen and discolored. There was weariness in the limbs, and a most distressing heat in the feet, especially at night, when the weight
of the bedclothes could not be borne. An erythematous
redness was often observed along the edges of the feet or
hands, and the sensibility, though generally heightened, was
occasionally diminished, so that they sometimes said that in
walking "they did not feel the ground." It was often localized in small patches, and not always accompanied by other
scorbutic symptoms. It was often combined, too, with low
fever or dysentery, and not unfrequently followed by des-
quamation of the epidermis, and sometimes by local gangrene.

In typhoid fever and scurvy, symptoms of a much less
pronounced character, but withal similar as to numbness,
formication, and hyperæsthesia, are sometimes seen at home.
In India a somewhat similar affection of the feet, called by

writers "**burning feet**," is mentioned by various writers* as being a most distressing disease of the Sepoys, and looked upon as being a sequela of rheumatism, and having its origin in the spinal cord.

All local treatment seemed unavailing in the Crimea, though stimulant and anodyne embrocations, hot and cold pediluvia, and shampooing appeared at times to assuage it. It disappeared as the general health and the state of the blood improved. Blisters were tried by some, but were manifestly injurious, and at times appeared to favor slough-ing, from the low vitality of the part.

It must be allowed that this affection, as it appeared in the East, bore a very close resemblance to the "**mal des pieds et des mains**," as it occurred in Paris in 1828. The writers of that period† tell us of the same pricking and formication of the feet and hands, the same streaking along their edges, the same alternating heightened and diminished sensibility, the œdema, dark patches on the limbs, and des-quamation of the epidermis, lancinating pains, and great heat of the parts increased at night, which were all so marked with us; but they had occasion **to** notice many severe symptoms which never showed themselves in our patients, as delirium, subsultus tendinum, severe gastric irri-tation, inflammation of mucous surfaces, (bronchitis, blenor-**rhagia, and** conjunctivitis,) affections of special senses, paralysis, and marasmus, sometimes followed by death. Nor in France were the local symptoms confined to the ex-**tremities,** but were sometimes extended to the face and trunk,

---

* See J. G. Malcolmson on Rheumatism and Burning Feet. **Mad-ras,** 1835.

† See Genest Arch. Generales des Med., t. xviii. **and xix.**; Char-don fils Rev. Med., t. iii.; Chomel, Chejoin, and Françoise **Jour.** Gen. de Med., t. **cv.**; Montault and Robert, **Do.,** t. **cvi. and cviii.**; Brous-sais An. de Med. Phys., t. xiv.; Dance Dic. **de** Med. Ozanam Hist. des Epidem. See also, on a similar affection which appeared in Padua, in 1762, Brugnatelli Jour. Physico-Medic.

as is also the Indian affection at times. In the cheiropo-
dalgia of France, many of these severe complications were
always present, and from the circumstances in which the
epidemic arose and spread, it is impossible to connect it
either with cold, wet, scurvy, or rheumatism. It continued
with them summer and winter. Among us it disappeared
in March with the fine weather and the improved diet. In
France it appeared chiefly in robust and plethoric persons,
whose symptoms were relieved by bleeding. With us it was
among those most "used up," in whom bleeding would have
been probably followed by gangrene. In France the disease
appears to have been a mixture of convulsive ergotism and
lead colic ; with us, an affection of nervous debility—a union
of weakness, cold, and scurvy. The "burning feet" of the
Sepoy resembles much more closely the acrodynia of France
than the Crimean affection did. The dropsical effusions,
spinal affections, fatal complications "ending in extensive
alterations in the structure of the viscera of all the cavities,"
and diffusion throughout the body, resemble more closely
what was seen in Paris than what was manifested in the
Crimea. In India anti-scorbutic remedies have most power
in overcoming it, as, I should say, they had also with us.

There are various traces of the occurrence of this affection
during the Peninsular war ; but it does not appear to have
attracted much notice from surgeons there. The disease
known as beriberi has, in some of its slighter forms, a resemblance to it, but has many symptoms of which we had no
experience.

In all these affections a depressed vitality and nervous excitement seemed to have been the chief causes of disease.
Dr. Tholozan says **he** observed on dissection **a peculiar**
"specific alteration" in the deep fatty tissues of the **hands,**
feet, and legs, especially in the borders of the feet, the pulps
of the toes, and the thenar and antithenar cushions, which,
he thinks, has not been as yet explained, and which he does
not connect either with scurvy, fever, dysentery or congela-

tion, and which, as he informed me, he believes to be the cause of the peculiar affection to which I have made reference above.*

I candidly confess, I was one of those who looked forward with foreboding to the chances of the plague appearing in our hospitals; but, providentially, we were spared

---

* "Le tissu graisseux de la plante des pieds, de la paume des mains, ou bien celui qui forme le coussinet sur lequel repose le ligament rotulien, ou bien les vesicules graisseuses situées contre le femur, au-dessus de l'articulation fémoro-tibiale, ont présenté 27 fois, **dans** les 79 autopsies, des altérations curieuses. Avec un état normal du derme et de l'aponévrose, on trouve les vésicules graisseuses sous-cutanées fortement injectées depuis le rouge clair jusqu'au rouge noir. La couleur jaunatre de la graisse a disparu derrière la forte injection. et même l'état ecchymotique de l'enveloppe celluleuse des vésicules. Ce n'est point une ecchymose sous-cutanée: c'est un état anatomique particulier, fort peu connu **du** tissu graisseux. Le tissu cellulo-fibreux intervesiculaire est normal et plutôt pâle, les cellules graisseuses sont très hyperémiées, et ces vésicules présentent quelquefois a leur surface un piqueté ecchymotique noirâtre. La graisse contenue dans les vesicules ne perait pas altérée.

"Cette lésion existe dans quelques cas en même temps dans les différentes régions indiques; souvent on ne la rencontre qu'à la plante des pieds ou au voisinage de l'articulation du genou; toujours elle est plus prononcée à la plante des pieds qu'a la paume des mains. Le tissu graisseux le plus altéré est celui qui avoisine le bord externe et le bord interne du pied, celui des éminences thénar et hypothénar. Quelquefois les vésicules graisseuses de la pulpe des doigts, ou des orteils ont présenté cette lésion, mais à un degré moindre. Au pied et à la main, la graisse située au-dessous de l'aponévrose n'est pas attaquée; le tissu graisseux sous-cutané ou profond des membres, ou des cavités splanchniques n'offre rien d'analogue à ces altérations. Il **ne m'a** pas été possible de saisir de relation entre cet état et le **scor**but, ou le typhus, ou la dyssenterie, ou les congélations. Il s'agit **là** d'une altération specifique non décrite, dont la valeur pathologique aurait besoin, pour être précisée, d'un plus grand nombre d'observations."
—*Recherches sur les Maladies de l'Armée d'Orient lues a l'Acad. de Med.*, *Sept. 30, 1856, par M. le Dr. Tholozan.*

this fearful invasion. Circumstances were certainly favorable to its outbreak; and, at one time, scurvy and malignant fever attained such a mastery as to wear many of the features of plague, both in the French hospitals at Constantinople and among the Russians at Odessa; but with us, the rapid amelioration which took place as the war proceeded made us less nervous about any development of it in our army.

A review, however superficial, of the medical annals of the war; of the hygienic causes, and local circumstances which led to the appearance and development of disease in our army, reiterates in trumpet-tones the same lesson—confirms and enforces the same conclusion—that the barometer of health rose and fell as external circumstances, favorable or injurious to health, were attended to or neglected. These circumstances were, in a great degree, under our own control, as will always be the case, whether in the camp or the city. This being so, it is surely the first duty of a government, as well as of a commander, to adopt every possible precaution which can guarantee the health and life of the army to which the honor, and even the safety, of the State are intrusted. By the adoption of judicious and enlightened means, disease, if it cannot be wholly banished from our camps, may yet be stripped of the deadly power which it now so destructively wields. Even wounds would become comparatively harmless, if all the vital powers, (the *vis medicatrix naturæ,*) possessed in full vigor and activity, were to put forth their mighty strength to sustain and restore the constitution. The soldier would have no enemy to fear but one he could see face to face—one whom the British soldier never fears—and thus the effectiveness of our army would be increased tenfold.

Any one who saw the two armies at Sebastopol: the ragged, gaunt, spectral-like figures guarding their fated trenches in the dreary winter of 1854–55, while the majority of their comrades lay in misery and pain in the wretched hospitals; and again witnessed the British army in the spring **of** 1856: every man in health and vigor—literally "full of lusty life," and actually "rejoicing in his strength,"— he who beheld that great contrast, and reflected how oversight and neglect were the causes of the one sad picture, and care, directed by knowledge and supported by energy, produced the other truly glorious one, would, whether actuated by principles of economy, humanity, or patriotism, **ever urge** his country **to** guard and preserve the *health* **of** the armies that defend it.

"Conserver les soldats," says Baudens, "transportés à grand peine, est le premier intérêt d'une nation qui fait une guerre lointaine; c'est aussi le meilleur gage d'un succès définitif. Les maladies tuent plus d'hommes que le fer et la poudre, et il est souvent facile de les prévenir par de simples précautions hygieniques."

8

CHAPTER IV.

THAT military surgery does not differ from the surgery of civil life, is an assertion which is true in letter, but not in spirit. As a science, surgery, wherever practiced, is one and indivisible; but as an art, it varies according to the peculiar nature of the injuries with which it has to deal, and with the circumstances in which it falls to be exercised. To the surgeon practicing in the camp, many accidents are presented which seldom or never come within the observation of the civil practitioner; while not a few of the cases which are daily treated in domestic life, rarely come under the charge of the military surgeon. The two classes of practitioners may be said to be engaged in separate departments of the same profession, which, though uniting occasionally, are yet tolerably distinct from one another.

The military surgeon during peace enters for a time into civil life; but during war he is called upon to exercise the very highest functions of his profession, and has little to do with the more trivial accidents which constitute the sum of a private practitioner's daily routine. His observation is undoubtedly restricted to a smaller variety of cases. He sees less than the civilian of the modifications which are impressed upon disease by age and sex; but in war he has a wider field for noticing the influence of external circumstances, of extremes of climate, of variations in food, work, and shelter on the same men, as well as the effects of mental

(86)

causes, as seen in the exultation of victory and in the prostration and dejection of defeat.

But though there may exist such distinctions between the spheres of the military and those of the civil surgeon, there is surely nothing in the exercise of their different callings which should create an antagonism between them. They are both members of the same priesthood, whose office it is to minister to suffering man, and the experiences collected by each should be willingly laid as common offerings on the altar of science.

To no class of professional men is a liberal education more important than to the army surgeon. To command that respect which is necessary for the right exercise of his official duties, he must be superior in general knowledge to his comrades. The many countries and varied climates to which he is sent, and the delicate positions in which his service often places him, demand the possession of an enlarged and well-stored mind; while the deep responsibility attached to the charge of such a number of valuable lives, and the necessity imposed by the absence of a "consultant" of deciding the most critical cases on his own unaided judgment, demand the firm self-reliance founded on clear knowledge as essential to any measure of success. Even amid the falling ranks, where he is exposed to as great danger as any, he must completely forget self, and give his whole mind to the condition of the sufferers around him; for often do his decisions, formed in a mere instant of time, settle for life or death the fate of the fellow-being before him. Then his powers of observation must be so well trained that he can discriminate between different diseases, whose types are mingled and masked by their union, as these are only seen in armies in the time of war.

The hardships incident to a soldier's life fall equally on the surgeon as upon his comrades; and, besides the dangers of battle and exposure, he runs the risk of those epidemic diseases which devastate armies, and which are the product of exciting causes, to which he has been as liable as any of those actually seized, and to the infection of which, when developed, he is ever exposed. In civil practice, on the other hand, a surgeon is not subjected to those predisposing and exciting causes of disease—cold, want of food and clothing, etc.—which cause its appearance among the mass of the population, nor does he remain exposed to its infection longer than is necessary to prescribe for his patient. The want of libraries for study and self-improvement are also drawbacks to the exercise of the profession in armies, of which the civilian has no experience.

The strict discipline which prevails in military hospitals gives the army surgeon some advantages over the civilian in the treatment of his cases. No interference from the ill-judged kindness of relatives, or from the headstrong willfulness of the patient himself, can occur. His opinion is a law from which there is no appeal, and thus fewer obstacles stand in the way of his giving a fair trial to remedies. He has, also, the advantages so often denied the civilian, of correcting or confirming his diagnosis and treatment by after-death examination—a point of the greatest moment. He can, in general, exercise his judgment also to the fullest without having his decision criticised by a host of ignorant censors, and thus the moot points in surgery can often be determined by him in a manner not permissible in civil life.

The greater uniformity in age, constitution, and external circumstances that is to be found among patients in the public services than among the mass of the population who enter civil hospitals, makes conclusions drawn from their treatment more reliable for future guidance in dealing with them, than any statistics derived from civil practice can be for general purposes.

But how different are the means of treating injury in the field and in civil life! The ample space, established routine, careful nursing, many comforts and appliances of a civil hospital contrast strongly with the temporary nature, hurried

extemporized inventions, and incomplete arrangements of a
military hospital in the field.

The influx of patients from the works of a besieging force,
or the shifting from place to place of an army during a campaign, makes the removal of the sick to the rear a necessity.
Then, as this transference has often to be accomplished by
means little adapted for the purpose, and at a period of the
treatment the worst fitted for its execution, the evil done is
often irreparable; so that injuries which might be completely
cured in stationary hospitals, have often to be relieved by
amputation, while others whose treatment might, under more
favorable circumstances, have afforded a fair prospect of success, are placed beyond recovery. From this it follows that
the military surgeon cannot always choose either his own
time or circumstances in performing his operations. He
must be content to do the best he can in the crisis, and thus
his experience has sometimes to be sacrificed to expediency.
His operations, too, often differ widely from the classic procedures of civil life. The adage, that "a good anatomist may
operate in any way," has often in him its illustration. The
object being to save as much as possible, compels him to tax
his ingenuity in order to take advantage of the eccentric
manner in which the ball has half accomplished the severance
of the limb, and to seize his flaps here and there where they
may be got; and thus, though the immediate result may not
appear so satisfactory, the final end is probably as effectively
secured. In the practice of field surgery, moreover, methods
of operating will often succeed which are not adapted for
civil practice. Thus, in the resection of joints which come
to be performed in the field, a comparatively small and simple
incision will enable the operator to remove the injured **parts**,
while in those cases in which the operation is commonly performed in civil life, a much larger and more complex incision
is generally required in order to permit of the extraction of
the enlarged, adherent, unbroken bone which has to be re-

moved, and perhaps to allow of the excision of part of the articular cavity at the same time.

As contrasted with the duties of the naval surgeon, those of the military surgeon are much more difficult. His patients are widely scattered, do not come so soon under his care when injured, are subjected to greater hardships, both immediately after being wounded and during treatment, than are the patients of the naval surgeon. "The sailor fights at home," while the unfortunate soldier has often much suffering to go through before he is admitted into hospital.

The soldier as a patient differs from the civilian in several well-marked points. In some respects he is a better patient, and in many respects he is a much worse one. Some of these points of distinction should always be borne in mind when estimating the success of surgery as practiced in the case of the one or the other.

Chosen when young from the mass of the population on account of his physical promise; selected with care during peace, with less discrimination during war, the soldier at starting is advantageously contrasted with the majority of the men of his own age. Chosen without any reference to his moral character, he is not uncommonly depraved and profligate in his habits, and has perhaps enlisted in the recklessness which succeeds to debauch, or as a last resource to save him from penury. We have thus, not unfrequently, two conditions meeting in the young recruit, both of which bear their own fruit in his future history—a tendency to indulge in vices which lead to disease, but a state of health in which disease has not been as yet established.

Taken from a domestic life in which he had possibly every liberty as to the disposal of his time, the formation of his habits, and the pursuit of his amusements, he is at once placed under the rigors of a discipline which soon becomes irksome. He enjoys little leisure, but is harassed by his unaccustomed, and, for a time at least, laborious duties. Nostalgia succeeds, and thus the period of acclimatization, as it

may be termed, becomes an ordeal so trying as in many instances **to** implant the germs of disease. The prejudicial
effects of this initiation will be the more sure, if the recruit
be launched into the real business of a war camp before his
constitution has had time to accommodate itself to the new
condition of things in which it is for the future to exist.
But if the young soldier get over this novitiate, then his
physical condition, during a time of peace at least, is undoubtedly favorable as contrasted with his fellow in civil life.
His food, which is well adapted for his use, is provided for
him regularly. He is systematically exercised. His hours
of labor and repose are carefully arranged, and he is at all
times liberally supplied with fresh air. The civilian, on the
other hand, though not subjected to the rough change of existence which the soldier has to undergo, is greatly less regular in his mode of life. He lives frequently in close streets
and airless dwellings. His food is irregular, varying with
the profits of his labor. He indulges without restraint when
he can afford it, and has to submit to privation afterward **to**
compensate for the excess.

In war, again, the soldier loses many of his advantages
over the civilian. The external circumstances which predispose to or generate disease are more numerous and vastly
more potent in his case than they ever are in civil life. The
exposure, the bad and irregular food, the deficient shelter,
the excessive fatigue, the unnatural excitement or depression
of victory or defeat, all tend to reduce him as much below
as he was formerly above the civilian in the scale of health.
He has, amid "the irregularities of war," opportunities for
licentiousness of which he is not slow to take advantage, and
his unquiet and exciting life is but too apt to occasion that
"debility of excess" which conceals a constitution weak to
resist injury, under an outward appearance of strength and
vigor. Thus it is, that as in civil life different trades produce different diseases, so a soldier's life, both in peace and

war, begets its own diseases, and secures exemption from others to which civilians are liable.

Morally as well as physically the sick soldier differs from the inmate of a civil hospital. If wounded, he received his injury in the discharge of his duty; if sick, in the fulfillment of praiseworthy service. His "honorable scars" recognize none of those causes referable to misconduct or stupid thoughtlessness, which so frequently make the civilian the inmate of a hospital. He has no fear like the civilian for the future, if incapacitated for further service, as he knows that his misfortune will entitle him to sustenance for the time to come, and that his country will regard him with gratitude.

When struck down by sickness, the soldier is, however, thrown more upon himself than the civilian, and this isolation must in his case act prejudicially on his recovery. He has no visits from sympathizing friends, as he lies on a sick bed, far from home, amid the selfish hardness of a camp. He is soon separated from his comrades, and placed among strangers gathered like himself from the accidents of the field, and he finds himself in circumstances where he has little to cheer but much to depress him. In the injuries to which he is exposed in war, he is more hardly dealt with than the civilian. The accidents which befall him equal in their severity the most terrible which occur in civil life. The effects produced by the massive round shot or ponderous shell are very like the crushing and tearing of machinery impelled by the resistless steam; so that, among the many assimilating effects of our railways and manufactories, one will evidently be, in course of time, the bringing of the surgery in civil hospitals more and more into conformity with that of war.

But, besides all that I have said as to those matters in which military and civil surgery are similar or disagree, and as to the contrast which exists on some points between the patients falling to be treated in either case, there are yet some circumstances in the late war to which I must allude, as they are peculiar in themselves, and have an especial bearing on its surgical annals.

A siege differs considerably from ordinary campaign work both in the description and mortality of the wounds to which it exposes the soldier.   The close proximity of the opposed batteries, the steady and deadly aim which can be obtained by the riflemen, the range so soon ascertained for cannon and mortar, the guns so carefully and accurately worked from the absence of hurry and from the daily practice of the gunners, all contribute to render the proportion of casualties higher and their severity greater in sieges than the injuries which attend a campaign in the field.   Wounds of the upper half of the body may be expected to be more common in a siege, from the lower parts being protected by the works, and shell wounds must also be of more frequent occurrence, from the larger employment of mortars in attacking or defending a city.*   The sudden sorties from the beleaguered garrison, the long and constant exposure to the enemy's fire while forming and guarding the trenches, all conduce to swell the number of those injured.

The health of the troops, moreover, does not maintain so high a standard when they are stationary, and want the wholesome animation which results from the change and stirring incidents of a moving campaign ; whence it follows that, on becoming inmates of the hospital, they are not so fit to stand active treatment, nor are they so "lively at recovery."

However, there is one advantage which a siege has over a campaign in the field, and it is a considerable one.   The hospitals, being more stationary, can be better arranged, and

---

* In the civil insurrections of Paris, they observed **the** greater frequency of wounds in the upper part of the body, and the consequently greater mortality among the revolted, who fired from windows and behind barricades, than among the soldiers, who occupied the open street.

placed so near the scene of conflict that the injured may be
more quickly succored.

During the late war, our army had not only to go through
the ordeal of great battles, but the prosecution of a siege
unequaled for its difficulties in the history of war—a siege
in which every obstacle and every trial was enhanced by the
stubborn resolution of a brave enemy and the frailty of our
own military preparation. The sorties were on a scale so
gigantic, and pushed so resolutely, as to occasion effects
little inferior to those of a pitched battle ; and the extraordinary length and active prosecution of the siege caused results resembling those of a constant battle several months in
duration. A few general engagements, and the casualties
of outpost service, make up the accidents of an ordinary
campaign ; but with us, day after day, and night after night,
kept up a constant strain, which was more exhausting to the
strength of the army than any other sort of warfare could
have been.

The majority of the recruits who joined the army early
in 1855, and who supplied many of the wounded of that
year, were far from being well chosen. They were selected
under a pressure, and were the contributions of a country
where the drag-net of the conscription is not used to inclose
the good as well as the bad, and where a soldier's life is not
in any honor or favor with the generality of the people.
Many of them were raw boys, ill conditioned, below the
standard age, undeveloped in body, unconfirmed in constitution, and hence without stamina or powers of endurance.
Often selected on account of their precocious growth, at
once launched into the turmoil, unwonted labor, and hardship of a siege in which the strength of full-grown men soon
failed, they were very quickly " used up. " Cholera or fever
speedily seized them, overtaxed as they were in work, and
unaccustomed to either the food or the exposure which fell
to them. The hospitals became filled with such unpromising patients, whose " wizened" look of premature age was

remarked by the most casual observer. If these unfortunate
boys were severely wounded, they almost invariably died, as
their weakly constitutions and overstrained powers could not
withstand "the ordeal of recovery." To them Hunter's
saying applied with peculiar force, that "their condition of
health did not bear disease." If they survived the first
effects of their injury, their convalescence was painfully prolonged, and the least imprudence produced a relapse. Their
ailments were seldom acute—their life-power was unequal to
its production; their nervous systems were shattered; and
that undefined but most fatal disease known as the "mal des
tranches" was soon set up. Depletory measures had soon
to be abandoned, and a more rational treatment, founded on
special symptoms and the observed effects of remedies, substituted for the conventional medication.

Again, several of the regiments which suffered most in
many of the assaults, and which consequently contributed
the greater number of the operative cases, were, either
wholly or in part, composed of men who had just returned
from prolonged service in India. Men so circumstanced
were but ill calculated to undergo the rigors of a Crimean
winter, or the hard work of the trenches, or yet the great
trial of a capital operation.

There was yet another element which demands attention,
when estimating the surgical records of the war. I refer to
the use of the new rifle, with its conical ball. The rifle used
by the Russians was little inferior in range or force to our
Minié, while its conical, deep-cupped ball was much heavier.
The great variety in form and weight which the balls used
by the belligerents presented will be seen by reference to the
table in the appendix, (E,) where the particular description
and weight of each are given. The greater precision in
aim, the immensely increased range, the peculiar shape, great
force, and unwonted motion imparted by the new rifles to
their conical balls have introduced into the prognosis of
gunshot wounds an element of the utmost importance. I

am not prepared to say whether the great destruction of the soft and hard tissues which these balls occasion results from their wedgelike shape, immense force and velocity, or the revolving motion, or from a combination of all these causes together; but of one thing I am convinced, that their use has changed the bearing of many points which fall to be considered by the surgeon in the field. The severity of the primary action on the part struck, and especially the aggravated evils which follow their wounds, combined to exercise a most prejudicial influence on the surgery of the war, to which due weight has never been given. Immense comminution of bone has been their most prominent effect. The amount of laceration of the soft parts seems to depend on the distance at which the missile is fired.

The wide-spread destruction of the bone often renders consolidation impossible, so that amputation has more frequently to be had recourse to, and the distance from the trunk at which that operation has to be performed being diminished by the same causes, the resulting mortality has been greatly increased. All who compared the dead of this with those of former wars, especially of Indian battles, were painfully struck with the greater disfigurement of the corpse caused by the conical than by any other species of ball.

But besides the more destructive nature of the small arms employed, cannons and mortars were used on both sides, of a caliber and range never before tried in any war. When **Paré** thought the cannon of his day so enormous and destructive, what can we say of those huge sea-service mortars and immense cannon used to defend and attack Sebastopol, compared with which those of the last century are as toys!* The fragments of our modern shells must **be as**

---

* "Truly," says Paré, "when I speak of the machines which the **ancients used for** assaulting men in combats and encounters, it appears to me as if I spoke of infants' toys in comparison with these, which, to speak literally, surpass in figure and cruelty the things which they thought the most cruel."

weighty as the whole projectile known to our forefathers,
and the grape which was so freely used in the East were
half as large as the round shot fired from the field guns in
the Peninsula. With us, every refinement in the art of
destruction was liberally practiced, so that "l'art de tuer les
hommes avec methode, et gloire," was, unhappily, never carried nearer perfection, though we may comfort ourselves with
the reflection of Percy, that this very perfection, "nous a
donné la même tâche et la même recompense dans l'art de
les conserver." "Les circonstances," says Briot, "qui con-
tribuent le plus à la destruction des hommes sont aussi celles
qui font decouvrir et developpent plus de moyens propres à
leur conservation."

Finally, if in war the surgeon sees much which is terrible,
much which taxes his feelings of humanity, and his regret
at the feebleness of his art, he has also the comforting conviction that nowhere is his beneficent mission so felt, nowhere is the saving power of his profession so fully exercised; so true is it that "chirurgery triumphs in armies
and in sieges. 'Tis there that its empire is owned; 'tis
there that its effects, and not words, express its eulogium."*

* Dionis, quoted by Sir George Ballingall.

9

CHAPTER V

THE "PECULIARITIES" OF GUNSHOT WOUNDS, AND THEIR GENERAL TREATMENT.

In saying that "there is a **peculiarity**, but no **mystery**, in **gunshot wounds**," John Bell has expressed the change of opinion which late times have brought about with regard to the nature of these injuries. It was the mysterious character ascribed by the old surgeons to wounds from so "devilish an engine" as a gun, which so long surrounded them with dread, and made incantations and charms the favorite resource in their treatment. The new philosophy has dispelled the mystery, but left us still to study the eccentricities which so often mark these injuries. The contused appearance and unavoidable sloughing of the walls of the ball's track, the little-suspected but serious destruction of deep parts, and the grave consequences which may ensue from such a wound appear to have been the circumstances that suggested the envenomed nature of gunpowder, and the cautery-like action of its projected ball, as well as the idea which prevailed, that in order to get quit of the injurious influences thus exerted on the wound, it was necessary to pour into it burning oil, or curious tinctures concocted from the most opposite and absurd ingredients, or to smear the part with nauseous grease and "charmed salves."

The description of the sensation caused by a gunshot wound in a fleshy part, usually given by the sufferer, **is, that** it resembles the effect of a smart blow from a supple cane. **Some,** however, feel as if a red-hot wire were passed through the part. The fracturing or splintering of a bone is always more painful than a flesh wound, and if a joint or

(98)

larger cavity be penetrated, the pain is still more acute, and the shock still greater—in most cases proportioned to the vitality of the part injured.

It is a very remarkable, though universally known fact, that when the mind is greatly engrossed by external objects —excited "'mid the current of the heady fight,"—severe wounds may be received without any consciousness on the part of the receiver. Whether the sensation may be so very slight as to be immediately obliterated by the tide of strong passions rushing through the mind of the combatant, or whether a reflex act of the mind be necessary for receiving a sensation—in common words, for perceiving the state of its companion, the body—I shall not attempt to discuss. But all military surgeons will confirm the statement of Hennen, when he says that "some men will have a limb carried off, or shattered to pieces by a cannon-ball, without exhibiting the slightest signs of mental or corporeal agitation— nay, without being *conscious* of it." I myself have known an officer who had both legs carried away, and who said that it was only when he attempted to rise, he became aware of the injury he had received; and very many who had suffered slighter wounds, have said that the trickling of blood along the skin was what first called their attention to their state.*

---

* This is a very curious and interesting subject to the physiologist, to all who study the marvelous interdependence of mind and body. What the exact province of each is, we are not in circumstances to determine, as we see all their operations carried on conjointly; but every one is aware that pain and sickness are greatly aggravated by the constant contemplation of them, and lightened by the mind looking elsewhere. The American Indians, whose stoicism has been so frequently extolled in song and story, were well aware of this last-mentioned law, for during the infliction of the most horrible tortures by the enemy, they sung the war-song of their tribe, and recounted the most glorious victories over their bitterest foes. Whether from philosophy or instinct, they directed the mind to the

The collapse and mental trepidation which frequently follow the infliction of a mortal wound in the trunk are, in many cases, most appalling. But although the presence or absence of this severe constitutional effect is useful as a diagnostic indication of the gravity of the injury, it is not entirely to be depended on, for the terror and amount of shock frequently depend as much on the nerve and frame of the sufferer as on the severity of the wound. The different effects produced on different persons by wounds in every respect alike are obvious to every one who has seen war, and call for the exercise of a most discriminating judgment on the part of the surgeon. Then, the period of collapse, which will, to some degree, occur in every case of a severe wound, varies greatly, which must determine whether immediate amputation be necessary, or whether it would be safe to delay it. The only other remark we make on this subject is, that the "commotion" succeeding gunshot wounds is greater when the lower extremities are injured than when the arms suffer; and this is more especially seen if the person be in an erect position when the injury is inflicted; which observation is consistent with the remark made by Chevalier, that the shock is always greater when the ball strikes a muscle in action than when it impinges against one which is relaxed.

The destruction inflicted by a ball depends on the dis-

---

most exciting and attractive topics, those best fitted to engross and absorb it, and to prevent it from looking at the wounds inflicted on the body, or listening to the taunts directed against the mind; and thus, if they did not actually prevent or nullify pain, they greatly lessened its intensity. We doubt not that the "noble army of martyrs" were often, through the same general law, enabled to rejoice even amid the flames, the mind being in a great degree absorbed by the contemplation of the glory awaiting them, revealed to them, as to the protomartyr, a brightness which the inner eye could behold, and thus they were almost

"laid asleep
In body, and became a living soul."

tance at which it is fired, the direction of its flight, its shape
and velocity, as well as on the nature of the part struck.
If fragments of metal are fired, as sometimes happened
during the sieges of the Peninsula, as well as in the civil
emeutes of Paris, and of which we had some experience in
the Crimea also, a very lacerated, irregular, and dangerous
wound may be caused.* A ball passing at great speed over
the surface of a limb may occasion a wound similar to that
made by a knife. But this action of a ball is rare.†

The great velocity, peculiar shape, and motion of the
conical ball give to its wounds a character considerably different from those which is present in wounds caused by a
round musket-ball. If fired at short range, and if it strike
a fleshy part, the conical ball produces, I think, less laceration of the soft parts than the old ball; but if the range be
great, and the part struck bony, with little covering of flesh,
as in the case of the hand or foot, then the tearing, especially at the place of exit, is greatly more marked.

I have not been able to satisfy myself in all cases, so
clearly as the description of authors would lead me to suppose I could, as to the characteristics which distinguish the
wound of entrance from that of exit. That the former is
more regular and less discolored than the latter, is true in
many cases, but that the lips of one wound are inverted,

---

* Hutin relates a case which occurred at the siege of Constantina,
where a nail was found fairly driven into the femoral artery; and
in the Burmese war, links of iron cable were fired by the enemy
from their cannon. Bullets, united together by wire so as to resemble "bar-shot," were at times used by the Russians in the Crimea,
and caused very irregular wounds.

† This, which is, I believe, the true state of the question, is opposed,
however, to Hunter's remark: "In this case (a ball passing with
velocity) a slough will be produced; but if it should pass with little
velocity, then there will be less sloughing, and the parts will, in
some degree, heal by the first intention, similar to those made by a
cutting instrument."—*Hunter's Works, by Palmer,* vol. iii. p. 559.

while those of the other are everted, has seldom been clearly marked to my observation. If the speed of the ball be great, and no bone have been struck, then there is little difference in either the size or discoloration of the wounds ; but if the flight of the projectile be so far spent as to be retarded by contact with the body, especially if it have encountered a bone or a strong aponeurosis, so that its speed is considerably diminished before it passes out of the body, then the wound of exit will considerably exceed in size that of entrance. This is especially true of conical balls. If, on the contrary, the ball be fired close at hand, so that its speed is not sensibly diminished by its passage through a limb, then the difference of size will be very small, and may even be in favor of the wound of entrance, as I had twice an opportunity of observing.

The usual action of a ball in proportioning the size of the two orifices is easily understood, when we consider that the part of entrance is supported by the whole thickness of the limb, while that of escape is quite unsustained, and therefore the more liable to be torn. Huguier has shown that the loss of substance which occurs at the place of entrance, and the flap-like tearing which takes place at the orifice of exit, form the best marks of recognition we possess, and that these characters can always be made out by examination of the clothes or accoutrements traversed in cases in which the supervention of inflammation has effaced them from the wound itself. The introduction but non-escape of a foreign body, as a piece of the breast-plate, belt, buckle, or part of the musket, etc., along with the ball, which alone passes out, or the flattening of the ball against a bone within, and its diameter being thus increased before it escapes, will all contribute to vary the relative characters of the orifices of the wound.*

---

* In Arnel's experiments, given in the *Journal Univer. de Med.* for 1830, it is shown that a ball, fired against a number of planks firmly

To the military surgeon, it is often of consequence to be able to conclude whether the two apertures in his patient's limb have been occasioned by one ball, which is thus seen to have passed out, or by two balls still imbedded in the limb, and to the medico-legal jurist, the knowledge of the marks which characterize the two wounds is of much moment.*

The action of a ball on the different tissues of the body may be, in a great measure, inferred from a consideration of the shape of the projectile, and the nature of the part struck. It carries away, as I before remarked, a piece of the skin at the place of entrance, and rends it where it escapes. The small plug of integument which is carried into the wound, Huguier tells us, can often be discovered there.†

The contusion which a ball causes in traversing **muscle**

---

bound together, causes a series of holes progressively increasing **in** size, so that a cone is formed by their union, whose base is represented by the last exit hole. M. Devergie's experiments on the same point, given in his communication to the Academy, go to prove this also. Velpeau and others have objected, but without good grounds, to the deductions drawn from the experiments being applied to the question.

* Between the opposite views held by Blandin and Dupuytren, the **opinions** of military surgeons and medical jurists have oscillated, **evidently from the fact that no constant relations exist between the** entrance and exit wounds. Velpeau, holding a middle view, concludes, with truth, "Dupuytren is wrong, and his antagonist is not right." The distance at which the gun is fired has most to do in determining their character, according to Devergie, who has himself, however, recorded a case which proves that the wound of entrance may be the larger, **even when the** gun is fired at a distance. Begin has given us the following valuable observation, with regard **to the** resulting cicatrixes. That of entrance, he says, is generally **white,** depressed, and often adherent to the underlying parts, **while that of** exit is only a sort of irregular spot, which does **not adhere to** the parts below, and **is** sometimes so indistinct as **to be** concealed in the folds of the skin. This difference he explains by the loss of substance sustained at the point of entrance.

† John Hunter speaks also of this piece of detached integument.

gives rise to one marked characteristic of gunshot wounds
—their healing only by suppuration and granulation. Occasionally an exception occurs to this rule. Thus, I have
seen a case in which a superficial wound of the gastrone-
mius was said to have healed without suppuration by the
fifth day, and in the records of a Sepoy regiment in India,
I find mention of even a deeper gunshot wound of the deltoid healing in the same way by first intention.

   Dr. Stewart, staff assistant surgeon, reports* a case of a
similar union, as having been observed by him during the
Kaffre war.   A Fingo received a pretty severe gunshot
wound of the muscles of the back, and union without suppuration took place.   Two things are necessary to produce
such a happy result: 1st, a most healthy and temperate
patient; and, 2d, the rapid flight of the ball.

   It is curious to notice how large a body may enter through
a muscle, and hide itself without producing any great wound.
Thus, I saw a case at Scutari, in which a piece of shell,
weighing nearly three pounds, was extracted from the hip
of a man wounded at the Alma, which had been overlooked
for a couple of months, and to which but a small opening
led.   Larrey gives a case in which a ball, weighing *five*
pounds, was extracted by him from the thigh of a soldier.
The presence of so large a body had not been detected by
the surgeon in charge, and the patient suffered no inconvenience from it beyond a feeling of weight in the limb.   Paillard mentions having heard M. Begin recount a case in
which a ball of *nine* pounds so buried itself for a time.
Hennen, too, mentions a case as having occurred at Sering-
apatam, in which a spent *twelve*-pound shot buried itself in
the thigh of an officer, and "so little appearance was there
of a body of such bulk, that he was brought to the camp,
where he soon expired, without any suspicion of the presence
of the ball till it was discovered on examination."   It is

---

* Unpublished Records of Medical Department.

more easy to understand how a large fragment of shell
should so conceal itself than a round shot, as, if its long
diameter corresponded with the run of the fibers of the
superficial muscles, and especially if the muscle was relaxed
at the time of contact, then a large piece might enter a muscular limb without causing an amount of injury proportioned
to the size of the body introduced. Such an instance occurred in the Crimea to a French soldier, of whose case
Baudens has given an account. A fragment of shell, weighing 2 kilog. 150 grammes, so completely buried itself in the
·thigh as almost to be invisible. The elasticity of the soft
parts doubtless assist in closing the opening by which such a
mass entered.

Baudens has made an observation which I am not aware
has been confirmed by any other, viz., that when the ball is
cut out from among the muscles, however early it may be
accomplished, it has a cellular envelope round it, which he
calls "kyste primatif," as contrasted with the "kyste defini-
tif," which forms its sac when it has been long inclosed in
the tissues.

Muscles which have been severely injured by ball are very
apt to become contracted during cure, if precautions are not
taken to prevent it. Of this most disagreeable result I have
seen a good many cases in the East.

On **tendons** a ball may cause little or no injury, especially
if they be relaxed at the moment they are struck. Their
toughness, elasticity, form, and mobility all help in protect-
ing them from being cut across, or pierced. A round ball
is often deflected by a strong aponeurosis like the "fascia
lata," particularly if it strike at an angle to the surface, **and**
at a period of its flight when the force is somewhat expended.
A conical ball, however, is seldom so turned.

It is on **bone** that the destructive effects of **a ball** become
most evident. (1) When its line of flight is very oblique,
and it is a flat bone against which it strikes, then it may be
thrown off, causing no other damage than depriving the **bone**

of its periosteum. When this occurs in the case of bones
of the head, much danger may subsequently ensue, as will
afterward be shown. Contused wounds of the long bones,
though seemingly of little moment at first, are sometimes
very serious in their results, not only from the separation of
the periosteum, and subsequent disease of the bone arising
from that source, but also from inflammation being set up in
the medullary canal. (2) A round ball may be flattened
against the shaft of a long bone, without causing any subsequent harm. This was often seen in India, where the match-
lock is used. (3) It may turn round a bone without breaking it. Thus, Chevalier records a case in which a ball,
entering at the lower part of the thigh, passed spirally
round the bone to the top of the limb, "comprehending
nearly the whole length of the bone in one circumvolution."
(4) A round ball, as is well known, may notch or partly
perforate a long bone without causing fracture, and pass off,
or remain in the medullary cavity, having passed through
the outer wall. This is, as can be easily understood, a most
dangerous accident. (5) If the force of propulsion be a
little greater, then the bone may be split longitudinally,
without being fractured across, as in a case related by Le-
veillé, and quoted by Malgaigne, in which an Austrian soldier
at Marengo was struck by a ball in the lower third of the
leg. He walked several miles to the rear, where he was seen,
and the wound thought to be very slight. A superficial
exfoliation of the bone was alone expected; however, his
symptoms became so serious that the leg had to be removed,
when it was found that, from the place where "the impression of the ball" existed, there proceeded several longitudinal and oblique clefts, which extended from the lower third
of the tibia up to near the head of the bone. (6) Into the
spongy heads of bones, and, more rarely, into their shafts, a
ball may be driven as into a plank of wood, without almost
any splintering, and become encysted there. (7) It may
pass through, causing a clean hole, of several of which

occurrences I will afterward relate cases; but the conical ball never acts in any of these ways, so far as I have seen. It is seldom split itself, but invariably splinters the bone against which it strikes to a greater or less degree, according to circumstances, and that in the direction of the bone's axis.   This tendency to splitting in the bone shows itself much more in a downward than in an upward direction, so that the destruction which such a ball will occasion will be greater when it strikes the upper than the lower end of a shaft.

All kinds of balls generally fracture and split the shaft of a bone if they strike it about its middle, but while a fracture with but little comminution results from the round ball, the conical ball—especially that which has a broad, deep cup in its base—splits and rends the bone so extensively that narrow fragments, many inches in length, are detached, and lesser portions are thrown in all directions, crosswise at the seat of fracture, and driven into the neighboring soft parts. It is the pressure of these fragments, as will be shown further on, which renders the fracture of long bones by the new ball so hopeless.*   I had many most interesting opportunities

---

* As instances of how great a difference it makes in the prognosis of cases whether a round or a conical ball has been the wounding agent, I may relate two cases, from a host of others. In the first instance, the ball entered on the external side of the ankle, near the tendo-achillis, and, passing forward and inward, lodged, as if in a piece of wood, in the lower end of the tibia, close over the ankle-joint. When the ball was removed, the bone was found not to have been split in any direction. A conical ball would have, to a certainty, opened the joint, and, in all probability, so split the tibia as to have necessitated amputation in the upper part of the leg. In another case, a round ball made a clean hole through one of the condyles of the femur, and did not split the bone; while, if a conical ball had struck the same part, it would have so cleft the bone that amputation in the middle of the femur would have been called for; whereas the removal of the limb at the knee-joint—a much less serious operation—sufficed in the case referred to.

of seeing the extraordinary manner in which the conical ball destroys bone in the way I refer to. I have never met with an instance in which such a ball, fired at whatever range, and striking at all perpendicularly on a long bone, has failed to traverse it and comminute it extensively.

From the comparatively little employment of the round ball during the late war, there were fewer illustrations of the splitting of balls on the edge of bone, as, for instance, on the edge of the tibia, or on the bridge of the nose, or on the humerus, than usually occur in a campaign. I do not believe that the conical ball, with its immense force of propulsion, could be so split. There is a case borrowed from Mr. Wall of the 38th, given later under wounds of the head, in which "a round rifle (?) ball" was thus split on the parietal bone, one-half entering and the other half going off externally, in a soldier of the 38th, wounded on the 8th September. Another somewhat similar case occurred in the 19th Regiment. It is by no means uncommon that a ball should be thus split on the head. Many examples of it occur in works on military surgery. No case clearly made out as one of splitting came under my own notice; but in one instance, a ball so changed in shape as to appear the section of one, was extracted from within the iliac fossa. Instances are on record in which balls have been split into three parts by the bones of the face and the trochanter major.

Although it cannot be for a moment doubted that balls may remain for a lifetime imbedded in bone, and cause little if any annoyance, yet it is equally certain that the most grievous results much more frequently arise from their presence in such situations. Of this, innumerable examples readily occur to any one who has seen many "veterans," or who has read much on the subject to which I refer. When speaking of wounds of the shoulder-joint, I will detail some cases which illustrate the pernicious action of balls left impacted in bone. Guthrie is very emphatic in his directions to remove balls so placed, and predicts the most disastrous consequences from the neglect of this measure. Malgaigne, after relating several cases in which balls have remained without causing harm, concludes thus: "It is necessary to mention these fortunate cases as evidence of the resources of nature, but they hardly serve to weaken the force of the prognosis when a ball cannot be extracted, or the essential indication of this sort of lesion—the extraction of the foreign body. This indication is, then, that of the first importance."

The nerves most commonly escape injury from a ball. If the missile has been rendered irregular in shape by previous contact with some hard substance, then it may do much damage to even the larger nerve trunks. Numbness, succeeded by pain in the extremity of a limb traversed by a ball, is not uncommon, and probably arises from the contusion or laceration of some chief nerve—the swelling and the pressure it occasions assisting to give rise to the subsequent uneasiness. The paralysis which succeeds the injury of a nerve may come on at once, or after an interval, and may or may not be accompanied with pain in the part itself, or in other regions connected with it by nervous communication. I have seen the hand several times waste when some of its nerves had been injured by a ball. In one case in particular, in which the ball had coursed up under the muscles on the external surface of the upper arm, this symptom was very marked.

Even though making all due allowance for the elasticity, strong coat, mobility, and form of the **arteries**, it is yet difficult to understand how they escape injury in gunshot wounds as they do. The rarity of primary hemorrhage on the field of battle has been long remarked, and yet how **often** do we meet with ball wounds apparently through the course of a great vessel!

The **veins** are more easily cut **than the arteries**, and primary hemorrhage, when it does occur, proceeds more commonly from them. Some vessels are more liable to injury

from balls **than** others. Thus, those firmly tied down, or
lying on bone, are more subject to damage than those loosely
reposing on the soft tissues. This remark applies especially
to two vessels: the femoral, as it passes over the brim of the
pelvis; and the popliteal, where it lies on the head of the
tibia. The lower parts of the ulnar, the radial, and the
facial, where it turns over the jaw, are subject to injury from
the same reason. An artery has not rarely been opened by
a spiculum of bone detached by a ball which had itself spared
the artery.

The **eccentric course** often pursued by balls has been a
frequent subject of remark, and though we had many most
striking instances of this, still I suspect we have had less of
it than occurred in the experience of former wars. The
conical ball seldom fails to take the shortest cut through a
cavity or limb, and it has at times been seen (as at the Alma)
to pass through the bodies of two men and lodge in that of
the third. But of the wanderings of the old round ball there
were many illustrations. I have known it enter above the
elbow, and be removed from the opposite axilla; and in
another case it entered the right hip, and was found in the
left popliteal space.* This "bizarrerie" in a ball's course
is accounted for by the deflecting action of tendons, apo-
neuroses, or processes of bone, or by the angle at which the
ball strikes, and the way in which, during certain positions
of the body, distant parts are placed in a line, as in the well-

---

* The surgeon of the 24th, when serving in India, mentions a case
in one of his reports, in which a ball entered below the angle of the
lower jaw, on the left side, and made its exit above the spine of the
right scapula, without injuring any important part; and M. Menière,
in his account of the Hôtel-Dieu during the "three days," tells us
of a ball which entered at the inner angle of the left eye, passed
downward, backward, and to the right side, under the base of the
cranium, and was removed above the right shoulder. The rapid
recovery, without a bad symptom, was no less wonderful in this
case than the direction taken by the missile was curious.

known case recorded by Hennen, in which a ball, entering the upper arm of a man ascending a scaling-ladder, was found half-way down the thigh of the opposite side. The fact of this wandering, however, is a peculiarity in gunshot wounds which often renders the discovery of the wounding agent difficult.

Foreign bodies, as pieces of cloth or part of the soldier's accoutrements, are often far more troublesome when introduced into a wound than the ball which occasioned their presence there. Innumerable and most heterogeneous have been the foreign bodies thus forced into wounds; but those which are capable of acting chemically as well as mechanically are the worst of all. Of these, lime, pieces of copper, etc. are the most frequently met with. Round lead balls are, perhaps, from their nature and shape, the least noxious of any, and are most likely to become encysted in the tissues.

Few questions connected with gunshot wounds have given rise to so much discussion and diversity of opinion as that with reference to the **extraction of balls**. For my own part I have seen enough to make me subscribe, with all sincerity, to Begin's precept, when he says in his communication to the Academy: "Selon moi l'indication de leur extraction est toujours presente, toujours le chirurgeon doit chercher a la remplir; mais il doit le faire avec la prudence et la measure que la raison conseille. S'il recussit, il aura beaucoup fait en faveur du blessé. S'il s'arréte devant l'impossibilité absolue ou devant la crainte de produire les lesions additionelles trop graves il aura encore satisfait aux principes de l'art; et quels que soient les resultats **de la** blessure il n'aura pas a se reprocher de les avoir **laisse** devenir funestes par son inertie."

If we examine into the opinions of surgeons **on** this point, we find that nearly all those who look upon the extraction of the ball as a matter of secondary importance are civilians, while military surgeons place great weight upon its accomplishment. The true way of putting the question is, not whether balls may remain in the body without causing annoyance, but whether they do so in so large a number of cases as to warrant non-interference. We must always remember that "science is not made up of exceptions," but is established by a collection of positive facts. Those who have studied gunshot wounds in the field, know full well how enormous is the irritability caused by the presence in a wound of a ball or other foreign body—how restless and irritable the patient is till it is removed—how prolonged the period of treatment is in the cases in which it is left; and how frequently the results are so distressing as to demand future interference, or condemn the unfortunate sufferer to a life of discomfort. As it is the surgeon's duty to treat his patients with reference to their future ease as well as to their present cure, so he should not try to bring about a healing of the wound which can be only temporary and fallacious, to the sacrifice of the efficiency of a limb and the future health of the body.

In this country we have not many opportunities of obtaining extensive information on the point as connected with the subsequent history of men with balls remaining unextracted, but such information is supplied from the Hôtel des Invalides of France, by M. Hutin, the chief surgeon to that magnificent establishment. He tells us, that while 4000 cases had been examined by him in five years, only twelve men presented themselves who suffered no inconvenience from unextracted balls, and the wounds of 200 continued to open and close continually till the foreign body had been removed. This epitome is of much value in estimating the question I am considering. In leaving the ball unextracted, we never know what evils may follow. The keeping open of the wound exposes the patients in the first place to all the dangers of a life in hospital, and the very elimination of the foreign body by suppuration, if it take place at all, necessitates a vast amount of annoyance. If it be a piece of shell or such like which is present, then its size will prevent its unaided extrusion, and the blocking up of the track, which it is so apt to occasion, may cause burrowing abscesses of a most destructive character.

Before a ball becomes **encysted**, it may set up grave inflammation, which will mat together and embarrass parts; press upon bone, and perhaps cause exfoliation; ulcerate blood-vessels, and so irritate nerves as to occasion affections as severe and fatal in their results as tetanus. It is somewhat remarkable, that in the wounded who came under my own care, two died of tetanus, in the very small number of instances—four or five at most—in which I could not find the ball. If this was a mere coincidence, it is the more curious. Gravitation and muscular action may so change the position of a ball, that from a harmless site it may be removed to one of much danger. It may thus work its way into a cavity, and cause fatal results.

But suppose the ball to become encysted in the first instance, what security have we that some very trivial circumstance (it may be a blow or even a deterioration in the health of the patient) may not set up irritation, inflammation, and suppuration in the cyst, and so come to set the ball free again to work harm in the economy? In any case its continued pressure gives rise to much uneasiness. The constant weight and weakness felt in the limb, the wandering pains, ascribed to rheumatism, from their aggravation by cold and damp, which attack even distant parts of the extremity, and the ever-present dread felt by the patient, if the ball be in close neighborhood to any vital organ, all unite to give much annoyance and discomfort.

The aversion which patients who have long carried unextracted balls express to have them removed, is not, as some would try to show, any proof of the slight annoyance they occasion, but simply indicates that they choose to suffer the discomfort rather than undergo what appears to them an

10*

uncertain and dangerous proceeding to free themselves of a
bearable inconvenience.

It seems, then, the teaching of experience, as it is of
common sense, that whether the question be viewed as one
bearing immediately or remotely on the result—on the cure
of the patient, in the proper acceptation of the term—then
we should, as soon as practicable, ascertain the position of
the ball, remove it along with any other foreign body which
may have been introduced with it, always supposing that by
such a proceeding we do not cause more serious mischief
than experience shows the presence and after-effects of the
ball can produce.

To **extract a** ball is in general not difficult. It is of much
consequence to proceed to its accomplishment before inflammation and swelling have come on, so as to close the wound.*
The great point to attend to undoubtedly is the fulfillment
of the rule, **which** is as old as Hippocrates, to place the
patient as nearly as possible in the same position as that he
occupied at the moment of injury—to put the same muscles
into action, and the angle which the parts form to one
another in the same relation; also, to place ourselves rela-
tively to him in a position to correspond as nearly as possible with the direction from which the ball came. By
considering the effect which bones or strong tendinous ex-
pansions may have had in deflecting the ball, or by paying
attention to what Guthrie calls the general "anatomy of the
whole circle of injury," and consulting the patient's own
ideas, which often convey to us most useful hints, we shall
in general succeed without much difficulty in discovering
the ball. An examination of the patient's clothes will show
us whether any part of them has been carried into and left
in the wound—whether the two holes seen in **the limb have**

---

* Percy adds another reason to encourage us in the early removal
of balls, when he says that men submit the more readily soon after
the receipt of the wound to the necessary incisions, before their
courage has been broken by pain and suppuration.

been caused by the same ball which has thus passed out, or by two balls which are still in; as well as whether the ball may not have carried in a *cul-de sac* of the clothes, and been withdrawn with it. If this be not attended to, very awkward mistakes may be made; as the mere correspondence in the direction of the two apertures, any more than their seeming want of relationship, cannot be taken as decisive in settling the matter. This point is well illustrated in the following instance related by an Indian surgeon : A wound was found below, and another above, the patella of a wounded man. The former had all the signs of the wound of entrance, and the latter those usually found at the place of exit of a ball. The opening of an abscess, which formed in the thigh a fortnight after, gave exit to a grape-shot, and it was found that the external condyle had been injured, and that each opening had been caused by a different ball.

In another instance, which occurred in the case of a soldier of the 40th Regiment in Cabul,* the ball appeared to have passed through the elbow-joint, and to have fractured the radius. There were two openings, having all the appearance of being those of entrance and exit; yet the ball was found and removed from the limb three weeks after. Such a mistake is most apt to arise when two balls have been fired together from the same gun, which happens not uncommonly in civil commotions, or when such fire-arms are used as the "espignol" of the Danes, from which a number of balls are fired in rapid succession, or when a cartridge, similar to that used during the Sleswick-Holstein war, is employed, in which two balls and a piece of lead are bound up together. One ball, too, it should be remembered, **may** make several openings. Thus, I have seen two in **the leg** and two in the hip, and also two in either thigh, occasioned in each case by one ball. **Dupuytren relates** a case in which, from its splitting, one ball made five **holes; and** the younger

---

* Unpublished Report.

Larrey saw at Antwerp six orifices caused in the same way. Sir Stephen Hammick mentions a case in which an aperture was found on either side of the chest of an officer shot in a duel. These corresponded both in position and character to those which would be occasioned by a ball that had traversed the chest; yet after death two balls were found in the body.

As showing the necessity of an early and careful search, as well as that we should never rely too much on the patient's statement, I may mention the following case : A soldier, wounded on the 18th June, came under my care in the general hospital. His right arm, which had been fractured compoundly, was greatly swollen at the time of admission. I was told, and accepted the story, that the accident had been caused by a piece of shell, to which species of injury the wound bore every resemblance, and that it had been removed by a surgeon in one of the trenches. At the earnest solicitation of the patient, I contented myself with applying the apparatus necessary to save the limb without minutely examining the wound. The injury turned out to be much masked, and to be greatly more severe than it at first appeared, the shaft of the humerus having been split into the joint. When removing the limb at the shoulder, some days after, to my great astonishment a large grape-shot dropped from among the muscles.* I before alluded to another case in which a piece of shell, weighing nearly three pounds, had remained concealed for two months without suspicion, from a like neglect of a proper examination.

It is well to remember also, in searching for balls, that they may have dropped out by the same aperture by which they entered, before we come to examine the case. Stromeyer has put us upon our guard against very curious

---

* I may, however, remark that this splitting upward of the head from the shaft is very rare. In general, the splitting ceases at the epiphysis.

errors, which he says he has seen made in cutting on the head of the fibula, and on a metatarsal bone for balls.

Sir Charles Bell has shown how the nerves may indicate to us the position of the ball. In one case he found it by pressing on the radial nerve, and so discovering that the ball lay behind it. "So when a ball has taken its course through the pelvis or across the shoulder, the defect of feeling in the extremity, being studied anatomically, will inform you of its course—that it has cut or is pressing on a certain trunk of nerve."

From all this, then, it is at the least very evident that we should not be too hasty in concluding that no ball remains in the limb, even although all the signs usually indicative of its having escaped are present; and also, that immediately before proceeding to take any steps for the removal of a ball, we should make certain of its position, remembering the rule laid down by Dupuytren—never to act upon information regarding the site of a ball obtained the day before, from the rapid manner in which they often shift from one spot to another.

The common **dressing forceps, if** long enough and fine enough in the handle, will, I believe, be found the most useful bullet extractor. That invented by Mr. Tuffnel, of Dublin, acted well in the few cases in which I tried it. Larrey employed **polypus forceps** in preference to anything else; but the inventions which have been made to accomplish this simple end are innumerable. To support the limb with the disengaged hand on the side opposite to that at which we introduce the forceps, is of much importance. If the course of the ball has been from above downward, **and** if it has approached at all near the surface, it should always be cut upon at the dependent part, by which **two** objects are secured—the removal is facilitated, **and an** opening for the pus is insured. If the wound **be large, as** it generally is from the conical ball, the finger forms the best probe, both to discover the ball and also to examine the state of the adjoining parts; otherwise a large gum-elastic bougie is our best resource. Causing the patient to move his limb, sometimes makes the site of the ball be felt by him, if not by **us**. Its position under a fascia, or in contact with a bone, would make us risk much in order to remove it.

The contentment of mind which results from the extraction assists recovery amazingly. The long continuance of the discharge, its gleety character, and the persistence of pain in the track almost always proceed from the presence of some foreign body—it may be a mere shred—in the wound. Chloroform is of inestimable service to us, both in the examination of wounds and in the removal of balls. All those voluntary muscular contractions which, although they are apt to interpose obstacles to our examination, were not presented to the entering ball, are done away with, and the severe pain which a prolonged examination and difficult extraction give rise to is avoided. We must, however, be careful to obtain from the patient all the information he can give us, before we bring him under the influence of the anesthetic.

The **inflammation** which ensues in a gunshot wound, shortly after its infliction, makes itself visible in the swelling and consequent eversion of the lips of both entrance and exit wounds, in the general tumefaction of the parts, and in the augmented pain. It was the fear of this inflammation strangulating the parts which gave rise to the exploded custom of scarifying the wound.* The swelling will differ

---

* Hunter expresses, with his usual clearness, the principles which should guide us in enlarging a wound, or "scarifying," as it was called. "No wound," he says, "let it be ever so small, should be made larger, except when preparatory to something else, **which** will imply a complicated wound, and which is to be treated accordingly. It should not be opened because it is a wound, but because there is something necessary to be done which cannot be executed unless the wound is enlarged. This is common surgery, and ought also to be military surgery respecting gunshot wounds."—*Hunter's Collected Works*, **vol. iii. p.** 549.

much in different regions and in different constitutions.   In
parts strongly bound down, in irritable tissues, in lax disten-
sible parts it will vary much, while, according as the patient
is of an inflammatory, lymphatic, or nervous temperament,
the effect will differ not a little.

The **constitutional fever** which sets in is generally proportioned to the importance of the part implicated, though
most anomalous exceptions do occur.   This fever will often
put on the characters of the endemic or epidemic fever; but
in war the tendency seems generally to be to a low typhoid
type, unless there be a decided local influence in action, as
that arising from paludal emanations.   With us, the symp-
tomatic fever must have been comparatively slight and eva-
nescent to what it was in the Peninsula.   The severe remedies
put in force by the surgeons of Wellington's army never
could have been employed by us.   That old soldiers, if
sober, are much less affected by this constitutional disturbance than others, is I think very observable.

The mitigation of the constitutional fever and of the local
inflammation, the prevention of all accumulations of matter,
by making judicious escapes for it, the relaxation of severed
muscular fiber, the application of light, unirritating dressings,
rest, and attention to the essential principles of all surgery,
comprise the general treatment which gunshot wounds usually
demand.   In the early stages cold may be of use locally—
even ice, as recommended by Baudens—and in wounds of
the hand and forearm irrigation is of the greatest service;
but when inflammation and suppuration are present, hot applications will always be found of most good.   Strict attention to the position of the limb is of great consequence, and
though in general it may be desirable, as in some instances
it is absolutely necessary, to restrict the diet, **yet** in those
cases in which much suppuration **is to be** expected, very
great latitude should be observed **with** reference to such a
rule.   Soldiers in war are commonly easily depressed, and
should not be fed too sparingly when admitted into hospital,

unless they suffer from a wound of the head, chest, or abdomen. Without placing too much faith on the happy effects which Malgaigne tells us the Russian wounded, treated in Paris in 1814, derived from a stimulant diet, as contrasted with the Prussians, French, and Austrians, still it is unquestionable that there is too much tendency to look on the common run of gunshot wounds as highly inflammatory, and to treat them accordingly. Velpeau's rule on this point agrees with his usual intelligent views, when he says he lays it down as a rule to remove his wounded and operated on as little as possible from their ordinary diet when they are hungry, and when there is no disturbance of the digestive and circulatory systems.

We found cod-liver oil of the greatest use in those cases in which the waste by discharge was great. A stream of lukewarm water, made to pass by gentle syringing from one opening to the other, forms one of the most useful methods of treating gunshot wounds Any shreds of cloth, clots of blood, pus, etc., which may be in the wound and sustain the suppuration, will thus be got rid of with very little disturbance to the parts. The addition of a little of Burnett's solution to the water thus used, or to the water-dressing, was useful at the same stage. Of the tonic and stimulant injections recommended by writers I had no experience ; but I have seen the French employ, with apparent advantage, a lotion composed of one part perchloride of iron and three of water in profusely suppurating wounds.

The extreme simplicity of the appliances and dressings employed during this war, and the nearly total absence of poultices, and such-like " cover-sluts," would, I think, have pleased Mr. Guthrie. The "stuffing in of great tents" was, I need not say, unknown ; and though we ascribed wondrous virtues to cold water, it was not on account of any "magical or unchristian" power which we supposed it to possess. Water-dressings, and the lightest possible bandaging consistent with the fulfillment of well-understood ends, were prevalent in our army, but not to the same extent among our allies, who have not yet given up the weighty pledgets of charpie and much fine linen, which so greatly astonish the English surgeon. The splints and other apparatus used partook of the simplicity of the rest of the treatment. Stiff bandages were too little used, if we accept the experience of the Sleswick-Holstein war; but the difficulty of always getting the necessary materials in the field is somewhat opposed to their use.

The state of the weather has got much to do with the rapid cure of gunshot, as of all other wounds. From a perusal of the medical records of regiments serving in the Colonies, it would, however, appear that hot weather, as in India, is, on the whole, more favorable than a cold climate.

Shell wounds, and grazes by **round shot,** are often followed by much injury, little suspected, but deeply seated, resulting, not unfrequently, in wide-spread sloughing of the soft parts. I cannot avoid relating the following case, although it did not occur in the Crimea, as it is a most excellent illustration not only of the great and, it may be, little suspected harm which may be occasioned by a round shot, but also because it is an instance of what would have been in former times set down to the wind of the ball. It is from the records of the medical department. Private John Conally was hit at Sadoolapore by a round shot on the outer **side of** the right arm and thorax. A blue mark alone was occasioned on the arm, and little or no mark was found on the chest. He died in twenty hours, without having rallied from the shock. The peritoneal cavity was found full of dark blood. The right lobe of the liver was torn into small pieces, "some of which were loose, and mixed with blood. There was no sign of inflammation in the peritoneum, and the other viscera were healthy." **Shell** does not comminute bone so much as a rifle-ball, but it tears the soft parts much more considerably. To refute the old myth concerning the effects of the wind of a passing ball, calls not even for pass-

ing mention in a work of modern times. All the cases of
this description of which I heard, were quite explicable on
the suppositions laid down by Vacher, in his memoir upon
this subject. Under wounds of the head, I have mentioned
a case (*Quin*) which would undoubtedly be set down of old
as having been so caused. There were many instances of
the very near approach and even slight contact of round
shot, without any inconvenience arising, further than might
be looked for from the unexpected and unwelcome vicinity of
such an intruder.

CHAPTER VI.

The Use of Chloroform in the Crimea—Primary and Secondary Hemorrhage from Gunshot Wounds—Tetanus—Gangrene—Erysipelas—Frostbite.

THE advantages derived from the use of anesthetics are perhaps more evident and more appreciated in the field than in civil practice. The many dreadful injuries which are presented to us in war, and the severe suffering which so often results from them, soon cause us fully to appreciate the benefits bestowed by such "pain-soothers."

The vast majority of the surgeons of the Eastern army were most enthusiastic in their anticipations of what chloroform was to accomplish. It was expected to revolutionize the whole art of surgery. Many operations, hitherto discarded, were now to be performed; and many, which the experience of the Peninsula said were necessary, were henceforth to be done away with.

In the British army chloroform was almost universally employed; but although the French also used it very extensively, as we learn from Baudens, still I do not think, from what I saw of its employment in their hospitals, that they had our confidence in it. Baudens tells us * that "they had no fatal accident to deplore from its use, although during the Eastern campaign chloroform was employed thirty thousand times, or more. In the Crimea alone," he continues, "it was administered to more than twenty thousand wounded, according to the calculations of M. Scrive."

In one division of our army it was not so commonly used

---

* Revue des Deux Mondes, Apr. 1857.

as in the others, from an aversion to it entertained by the principal medical officer of the division—a gentleman of very extensive experience. The only case in which, with any show of fairness, fatal consequences could be said to have followed its use, occurred in the division referred to. The patient, a man of thirty-two years of age, belonged to the 62d Regiment, and was about to have a finger removed. The chloroform was administered on a handkerchief, as he sat in a chair. Death was sudden; and artificial respiration, which was the means of resuscitation employed, failed to restore him. No pathological condition sufficient to account for death was found post-mortem. Some five or six other cases were brought forward, by the small body of surgeons who were suspicious of the action of chloroform, as having ended fatally from its effects; but in none of these could, I think, the least pretext be found for the imputation, further than that the anesthetic had been administered at some period previous to death. A man who had been dreadfully mutilated, and who had lost much blood, died shortly after having his thigh removed **high up.** Chloroform had been used, and to it was ascribed the fatal issue. Death, twenty or thirty hours after a capital operation, rendered necessary by the most dreadful injuries, must be attributed to the chloroform, and so on, and no note taken of the effects of severe injury, *plus* a capital operation, in shattering the already enfeebled powers! Death occurring under such circumstances, when no chloroform was employed, would not be thought to demand any special explanation, nor does the fact that the injury was occasioned by a round shot introduce any new element into the calculation.

The objections made to the use of chloroform were restricted to two classes of cases—trivial accidents, in which it was thought unnecessary to run the risk of giving it, and amputations of the thigh, in which a fatal accession of shock was feared. However this may be, it certainly shows the little practical force of these objections that, while with every indulgence in the interpretation of the law *post hoc*, etc., only some half-dozen cases could be obtained throughout the whole army to illustrate the pernicious effects of this agent, and that, too, when thousands upon thousands had been submitted to its action, and hundreds of surgeons **of** equal experience to the objectors were ready to record their unqualified opinion in its favor, as well as their gratitude for its benefits. For my own part, I never had reason, for one moment, to doubt the unfailing good and universal applicability of chloroform in gunshot injuries, *if properly administered*. I most conscientiously believe that its use in our army directly saved very many lives—that many operations necessary for this end were performed by its assistance which could not otherwise have been attempted—that these operations were more successfully, because more carefully, executed—that life was often saved even by the avoidance of pain—the *morale* of the wounded better sustained, and the courage and comfort of the surgeon increased. I think I have seen enough of its effects to conclude that, if its action is not carried beyond the stage necessary **for operation, it** does not increase the depression which results from injury, but that, on the contrary, it in many instances supports the strength under operation. Its usefulness is seen in nothing more than when, by its employment, we perform operations close upon the receipt of injury, and thereby, if not entirely, at least in a great degree, are able to ward off that *embranlement* of the nervous system which is otherwise sure to follow, and whose nature we know only by its dire effects.

To men who had lost much blood it had, of course, to be administered with great care, from the rapidity of its absorption in such persons; but if we do not act on broader principles in its exhibition than reckoning **the** number of drops which have been employed, or the part of the nervous system which we **may presume to be at the** time engaged, then we must expect disastrous results. It is difficult to see how its use could favor secondary hemorrhage after op-

11*

eration, as some said it did; but it is, on the contrary, easy to understand how the opposite result might follow. That purulent absorption should prevail among men so broken in health as our men were, need not be explained by the employment of chloroform; and that ice would prove more useful, in the slighter operative cases in field practice, few will be disposed to admit, either on the ground of time, efficiency, or opportunity. To Deputy Inspector-General Taylor we owe the practical observation, that chloroform appears to act more efficiently when administered in the open air.

In the prolonged searches which are sometimes necessary for the extraction of foreign bodies, chloroform is useful, not only in preventing pain, but also in restraining muscular contractions, by which obstacles are thrown in the way of our extraction which did not oppose themselves to the introduction of the body. Then much is gained in field practice by the mere avoidance of the patient's screams when undergoing operation, as it frequently happens that but a thin partition, a blanket or a few planks, intervene between him who is being operated upon and those who wait to undergo a like trial. Thus when, as after a general engagement, a vast number of men come in quick succession to be subjected to operation, it is a point of great importance to save them from the depression and dread which the screams and groans of their comrades necessarily produce in them.

It is therefore my clear conviction that the experience of the late war, as regards chloroform, is unequivocally favorable; that it has shown that chloroform, both directly and indirectly, saves life; that it abates a vast amount of suffering; that its use is as plainly indicated in gunshot as in other wounds; and that, if administered with equal care, it matters not whether the operation about to be performed be necessitated by a gunshot wound, or by any of the accidents which occur in civil life.

Hemorrhage was in the olden time the great bugbear of the military surgeon, and that against which his field arrangements were chiefly directed. It is not now, however, so much feared, from its being well known not to be of so frequent occurrence on the field, and the means of arresting it being better understood. Blandin, in his communication to the Academy, says that his observation of gunshot wounds led him to believe that primary bleeding always takes place if a vessel of any size is injured, but that it is soon spontaneously arrested by an action similar to that which takes place when a limb is torn off. Sanson repeats the remark as to the constant presence of hemorrhage to some extent at the moment of injury. Guthrie did good service to surgery by showing how small a force can obstruct a vessel of the first order. He thereby gave courage and confidence to both surgeon and patient.

It has been the experience of most wars, certainly of the late one, that tourniquets are of little use on the battle-field; for though it is unquestionable that a large number of the dead sink from hemorrhage, still it would be impossible, amid the turmoil and danger of the fight, to rescue them in time, the nature of the wounds in most **of these** cases causing death very rapidly.* A great artery is shot through, and in a moment the heart has emptied itself by the wound. It would be an experiment of some danger, but of much interest, as bearing on this question, to examine the bodies of the slain immediately after a battle, and carefully record the apparent cause of death in each case.†

---

* Although this is true as a general rule, yet both Larrey and Colles relate instances in which, by prompt assistance, death **was** prevented in wounds opening the carotid artery.

† The only mention I have been able to meet with, in the **records** of the medical department, of the causes of death on **the field, is in** a report from the surgeon of the 41st, when serving **in Cabul.** He mentions that of four men **who were** killed outright, three were wounded through the chest, **and one through the** head. After the contests in Paris in 1830, Ménière tells **us that** it was a common observation at the Morgue, to which **the dead** were carried, that the greatest number had been shot through **the** chest.

I before remarked how curiously arteries escape injury from a ball passing through a limb. In the course of the femoral vessels this is very commonly seen. Through the axilla, through the neck, out and in behind the angles of the jaw, between the bones of the forearm, and even of the leg, balls of various sizes and shapes pass without injury to the vessels. Thus, the neck has suffered severe injury many times, and yet very few deaths appear in the returns from these wounds. I have never myself seen any case in which a bullet has passed harmlessly between a large artery and its vein, but many such cases are on record.

The following may be mentioned as instances of narrow escapes : A soldier was wounded at Inkerman, by a ball **which entered** through the right cheek, and escaped behind the angle of the opposite jaw, tearing the parts in such a manner that the great vessels were plainly seen, bare and pulsating, in the wound. Three weeks after admission into hospital he was discharged, never having had a bad symptom. A soldier of the Buffs was wounded in June, 1855, by a rifle-ball, which struck him in the nape of the neck. It passed forward round the right side of neck, going deeply through the tissues ; turning up under the angle of the inferior maxilla, it fractured the superior maxillary and malar bones, destroyed the eye, and escaped, killing a man who was sitting beside him. This patient made a rapid recovery. **A French** soldier at the Alma was struck obliquely by a **rifle-ball, near to but** outside the right nipple ; the ball passed seemingly quite through the vessels and nerves in the axilla, and **escaped** behind. His **cure** was rapid and uninterrupted. Another Frenchman was struck in the trenches by a ball, a little below the middle of the right clavicle. The ball escaped behind, breaking off the upper third of the posterior border of the scapula, and yet he re**covered** perfectly, without any **bleeding taking** place. Endless numbers of similar cases are presented to us in military hospitals.

A considerable artery may be fairly cut across, and give no further trouble, beyond the **first gush** of blood which takes place at the moment of injury. In such cases, the vessel contracts and closes itself. If only half divided, as it is apt to be by shell, or by the quick passage of a ball, then the hemorrhage will be, in all probability, fatal. The best example, perhaps, on record of the former result, is that mentioned by Larrey. A soldier, struck on the lower third of the thigh by a ball, suffered one severe hemorrhage, which was never repeated. The limb became cold, the popliteal ceased to beat, and the ends of the divided femoral could be felt retracted when the finger was placed in the wound. This man recovered perfectly. The younger Larrey records a very curious case from the wounded at the siege of Antwerp. A shell passed between a man's thighs, and, destroying the soft parts, divided both femorals; yet there was no hemorrhage, although the pulsation continued in the upper ends of the vessels to within a few lines of their extremities.

The speed of the ball at the moment when it comes in contact with an artery has a good deal to do with the injury it inflicts. If it be in full flight it may so cut open the vessel as to allow of instantaneous hemorrhage; whereas, if its speed be much diminished, the contusion it occasions opposes **immediate,** but favors secondary bleeding.

Primary hemorrhage may take place either instantaneously **on** the receipt of a wound, or after a little time, when the faintness resulting from the accident has gone off. I have already referred to some instances in which the former is liable to occur. In wounds of the face, too, this instantaneous bleeding is very usual.

Some cases occurred in the Crimea of the **well-known** fact that limbs may be carried away, and their **arteries** hang loosely from the shattered stump, witho**ut bleeding.** Two came under my own notice, in which legs **were** carried away by round shot, and no hemorrhage took place, though both men died subsequently from other causes. This spontaneous

cessation of hemorrhage is perhaps most commonly seen in
the upper arm.

The returns fail to inform us of the number of cases,
either absolutely or proportionately to the whole number of
wounds, in which secondary hemorrhage took place during
the war. Although I have no figures to which I can refer
as corroborating the statement, yet I am inclined to think
that the proportion of cases in which serious bleeding did
take place is higher than that set down by Mr. Guthrie.
The distinction drawn by Dr. John Thomson between
secondary hemorrhage, proceeding from sloughing, ulceration, and excited arterial action as it occurs at different
stages of treatment, is a good one. That which takes place
after twenty-four hours and up to the tenth day being
usually due to sloughing, resulting directly from the injury,
should always have the term "intermediary" applied to it;
and the bleeding which proceeds from morbid action, such
as ulceration attacking the part, and which takes place at a
later period, would be more appropriately called "par excel-
lence" secondary. Hemorrhage should thus be distinguished
into three periods: **primary**, occurring within twenty-four
hours; **intermediary**, between that and the tenth day; and
secondary, that which takes place at a later date. More
precision would be given to our language on this important
subject, by such a distinction being always recognized.

The period at which consecutive bleeding is most apt to
take place has been variously estimated. Guthrie sets it
down as occurring from the eighth to the twentieth day,
Dupuytren from the tenth to the twentieth, Hennen from the
fifth to **the eleventh, and** Roux from the sixth to the twentieth. In the cases I have myself observed, it has taken
place between the fifth and twenty-fifth days, and by a curi-
ous coincidence, it has appeared in the majority on the fifteenth after the receipt of the wound. In one case, a wound
without fracture of the thigh, it was said to have taken place

as late as the seventh week, and that when no gangrene or apparent ulceration was present.

Consecutive hemorrhage may occur from very insignificant vessels, and be arrested by simple means; but when it takes place from a large arterial trunk, it is an accident of the most serious importance.* With us, in particular, such effusions were causes of extreme anxiety, as the deteriorated state of the health of our patients made such an accident peculiarly disastrous. Their strength could not withstand such a drain, and the scurvy made their blood so thin and effusible that they were liable to great loss of blood, not by vigorous hemorrhages, but by slow, though not less destructive discharges. From this it can be understood that in the Crimea many of the time-honored remedies for hemorrhage, such as venesection, starving, etc., were entirely discarded, and replaced most generally by their opposites. Tonics, as quinine and iron, were the remedies most wanted; and as to styptics given internally, they always appeared to me to be mere farces, except in so far as they acted as general tonics.

The more useful prophylactics to such consecutive hemorrhages, such as quiet of mind and perfect rest of the wounded part, are not always attainable in field practice, especially when the necessity of removing patients occurs so

---

* The following is a very interesting case of secondary hemorrhage caused by the limited ulceration of a large artery, which is related by Dr. Scott of the 32d, in a report existing in the archives of the medical department: "Private John Hodgson, aged thirty-one, was struck by a ball at Mooltan, about a line anterior to the left carotid artery, below where it divides into the external and internal, and, passing through the œsophagus, escaped at a point corresponding to its entrance. No unfavorable symptom appeared for nine days, when a fit of coughing came on, and blood issued from both the mouth and the wounds, and the patient instantly expired. The right carotid had been grazed at its bifurcation, and a piece of it, about the size of a small pea, and including all its coats, had sphacelated, and, giving way, caused death before assistance could be got."

frequently. It is of course impossible altogether to avoid such movements **during** war, but it is most unfortunate that they fall so often to be executed at the very period when they become most dangerous. No man, at all severely wounded by gunshot, can be considered safe from hemorrhage **till** his wound is closed, but yet, after twenty-five days, the danger may be said to be in a great measure overcome. In reference to this point a siege has an advantage over an open campaign, from the greater fixedness of the hospitals, and the less frequent moving.

Hemorrhage occurring early was universally treated by the rule laid down by Bell and Guthrie, of tying both ends **of the** bleeding vessel. When, however, the bleeding appears at a late date, when the limb is much swollen, its tissues infiltrated, matted together, and disorganized, it is by no means an easy thing to follow this practice. The difficulty is perhaps greatest in wounds of the calf of the leg, where the muscles are much developed, when the posterior tibial has repeatedly bled, the wound large and irregular, the contusion severe, and the blood **welling out** from among the disorganized tissues in no collected stream. The rules and precepts laid down in books about the appearance of the vessel and the orifice, about the mode of passing a probe toward it from the surface, and the best way of cutting so as to fall upon the vessel, are all worse than useless, as they **lead us to** expect guides where there are none but those which watchful **eyes** and careful incisions afford.

From the **results of** several cases which fell under my observation **in the** East, **I have** reason to believe in the soundness of **the views** lately put forth by Nélaton, in opposition to the long-credited opinion of Dupuytren, as to the unsound state of the artery in suppurating wounds. I feel pretty sure that the vessel will, in most **cases, bear a** ligature **for** a sufficient time to fulfill the end we have in view in its application. It will be necessary to attach it with caution, **to employ no more** force than is absolutely necessary, and we may expect it to separate, as Nélaton shows, before the usual time, yet it will continue attached sufficiently **long** to close the vessel, if we do not keep pulling at it so as to tear it away prematurely. It requires but a small force to oppose the blood-impulse, and that the vessel will commonly stand, if carefully handled.

The French, although generally applying the ligature at the seat of injury in **primary hemorrhage**, perform Anel's operation when the bleeding appears late. The teaching of Dupuytren and Roux has done much to prevent "the English practice" being so fully followed as it is with us. *

---

* M. Roux, in the second volume of his recently published posthumous works, thus sums up his experience on secondary hemorrhage from gunshot wounds. It proceeds, he says, from (1) separation **of** the eschar; (2) from injury by fractured bones; (3) from the capillaries caused by general feebleness in the patient; (4) hemorrhage from the erosion or tearing of a vessel appears later than that arising from the separation of the eschar, the **one** appearing about the eighth **or** tenth day, and the other from the fifteenth to the twentieth; (5) hemorrhage arising from the **tearing** of **a vessel, and** especially that which **accompanies compound fracture, is more common** in wounds of the **thigh than any other;** (6) **whatever be its** cause, the manifestation of **the** bleeding is very uncertain, being sometimes preceded by symptoms which announce its approach, and **sometimes** giving no indications of **its** coming—sometimes it appears **in large** quantities, and **very** suddenly, while at other **times** it appears **in small quantities, and** will often recur if no interference be had recourse **to;** (7) sometimes the bleeding takes place within the limb, where it forms a sort of false aneurism, but at other times it flows freely outward; (8) when the bleeding vessel can be got at, we should ligature it, or the trunk from which it proceeds; (9) here, **as** in the case of all traumatic hemorrhages, primary or secondary, **it** is best to tie the vessel in the wound, above and below **the place of** injury; in general, however, it will be necessary **to** ligature the vessel at a distance on the distal side of the wound, **after** the methods of Anel or Hunter, because the difficulties of finding it are great, and its state in the wound will not allow **of a** ligature being applied to it there.

Sanson, again, (Des Hæmorrhagies Traumatiques,) concludes thus:

Anel's operation is undoubtedly the best in one class of cases dwelt upon by Dupuytren, in which hemorrhage arises from the tearing of an artery in a simple fracture. The ligature of the main vessel commonly succeeds in these cases, while to find the bleeding vessel is most difficult, and to expose the seat of fracture to the air is a risk greater than should be encountered.

There are, unquestionably, some situations where it is impossible to get at the wounded vessel, especially when the bleeding has taken place at a late date. The deep branches of the carotid afford, perhaps, the most patent example. In a case of this sort, in which the deep temporal and internal maxillary were wounded, in a Russian admitted into the general hospital after the assault in September, Mr. Maunder tied the carotid to arrest the bleeding, which had recurred several times, notwithstanding pressure. The ligature of the main vessel commanded the hemorrhage, although the patient subsequently died of exhaustion.

Secondary hemorrhage may appear at a very late date from ulceration, set up by the pressure of a fragment of

---

"A ligature applied to the two ends of a divided artery is the surest method of arresting the bleeding, and to prevent a return. But we do not think, after the example of the English surgeons, that it should be put in force in all cases, and whatever be the situation of the artery, from the risk of causing great destruction, violent inflammations, and long suppurations. We often meet with wounds attended with hemorrhage, or false primitive aneurisms, in which it is difficult or impossible to determine which is the divided vessel. In other cases we recognize the source of the bleeding, but it is situated too deeply for us, without causing grave injury, to find and tie the artery above and below the wound. We are thus compelled to ligature this artery, or at least the trunk from which it proceeds, between the heart and the wound, but at a considerable distance from the latter. It is true that traumatic hemorrhages are much less favorable than aneurisms, properly so called, to the success of Anel's method. But it is a necessity in a way, and besides, we can, if the method of Anel fails, have recourse at a later period to the ligature of the two ends in those cases in which it is possible."

bone pressing upon the vessel. The ulcerative process in these cases is sometimes very slow.

The following case is interesting, and conveys much instruction as to the value of the different places in which to apply the ligature: A Russian boy who had sustained a compound fracture of the leg at Inkerman, from gunshot, was received into the French hospital at Pera a few days afterward. On the fifteenth day from the date of injury, profuse hemorrhage took place from both openings. Pressure failed to arrest it. The popliteal was tied the same day, according to the method recommended by M. Robert, viz., on the inner side of the limb, between the vastus and hamstring muscles. The foot remained very cold for four days, and then violent reaction set in; and on the eighth day from the ligature of the main vessel hemorrhage recurred, both from the original wound and the incision of ligature. Pressure was again tried in vain. The superficial femoral was next ligatured, on the tenth day from the deligation of the popliteal. Four days afterward the bleeding returned from the wound, and pressure then seemed to check it. The ligature separated from the femoral on the twelfth day after its being applied, and the third day, *i.e.* the twenty-fifth day from the first occurrence of the hemorrhage, bleeding having again **set** in from the wound, the limb was amputated high in the thigh, and the unfortunate patient ultimately recovered. Would Mr. Guthrie not have saved this boy's limb, and the surgeons much trouble?

In gunshot wounds of regions where the vascular communications are at all free, the ligature of the main trunk for consecutive bleeding cannot often be of any use, as **it is** seldom possible to be sure that **the** hemorrhage proceeds from the main vessel, nor yet can we by such an operation cut off the collateral circulation. **If the** source of the hemorrhage could be certainly ascertained, and if pressure could be applied to the lower portion of the divided vessel at the same time, then we might reasonably hope to arrest

the bleeding by tying the main artery; but the mere placing
of a ligature on the proximal side can give no security
against the continuance of the bleeding. If the sloughing
preceding and accompanying the bleeding be extensive, and
situated in a muscular and vascular part like the calf of the
leg, and if the hemorrhage has continued notwithstanding
the employment of means applied locally, I should never
hesitate between amputation and ligature of the main trunk,
but have instant recourse to the former, as being the only
reliable and satisfactory proceeding.* The following may
be taken as a good example of a class of cases frequently
occurring: M'Gartland, a soldier of the 38th Regiment, an
unhealthy man, who was still suffering from the effects of
scurvy and fever, was shot from the outside and behind, forward and inward through the left leg, on the 18th of June.
The fibula was broken, and the edge of the tibia was injured.
He walked to the rear without assistance. On admission
into the hospital, the limb was greatly swollen. This swelling, by appropriate means, very much diminished in a few
days. On the fifth day arterial bleeding, to a limited extent,
took place from both openings. Recalling a case put on
record by Mr. Butcher, of Dublin, of a wound of the post
tibial, I determined on trying the effects of well-applied

---

* I may note the following figures, in passing, as a small contribu-
tion to the statistics of this question: The French, in one hospital
at Constantinople, ligatured the femoral at a distance from the wound
for secondary hemorrhage seven times, and all failed.   The subcla-
vian was ligatured under like circumstances once, and it succeeded.
I have found the detailed report of only four cases in which the main
vessel was tied in India.   The ligature was applied twice to the
femoral, once to the brachial, and once to the radial.   It succeeded
in arresting the hemorrhage in three cases; one femoral failed.   Dupuytren ligatured the femoral several times for bleeding from the
calf, but with what result it is impossible always to make out.
S. Cooper, while he once successfully took up the popliteal for
secondary hemorrhage from a wound of the posterior tibial, strongly
reprehends the practice as a general rule.

pressure along the course of the popliteal and in the wound, combined with cold and elevation. The limb was also fixed on a splint. The object of the pressure on the main vessel was to diminish, not arrest, the flow of blood through it. On the eighth day there was again some oozing. Pus had accumulated among the muscles of the calf, and required incision for its evacuation. On the ninth day a pulsating tumor was observed on the external aspect of the leg, "the consecutive false aneurism" of Foubert, and next day the bleeding returned from both wounds. I wished then to cut down and tie the vessel in the wound, but a consultation decided on waiting a little longer, in the hope that the bleeding might not return. On the night of the eleventh day most profuse hemorrhage recurred. The attendant, though strictly enjoined to tighten the tourniquet, failed to do so, but the necessary steps to arrest the bleeding were taken by the officer on duty. Next morning, when I first heard of the occurrence, I found the patient blanched, cold, and nearly pulseless. A consultation decided that the state of the parts made the securing of the vessel in the wound very problematical, and that, as the limb would not recover if the main artery was taken up, amputation must be performed so soon as the patient had rallied. When reaction had fairly taken place, I amputated the limb. The removed parts were much engorged, sloughed, and disorganized. The anterior tibial was found to have been opened for about an inch shortly after its origin, and on it was formed the aneurism, which had a communication with both orifices of the wound.

In all such cases the second bleeding should determine active interference. I say the second bleeding, as it **very** often happens that when hemorrhage has taken place once, even to a considerable extent, and evidently from a vessel of large caliber, it never recurs. Many most striking instances of this have come under my notice. But though more than even this is true, and that frequently blood thrown out repeatedly is spontaneously arrested, still the great preponderance of cases in which it recurs in dangerous repetitions and quantities, as in the above instance, should cause us, I believe, to interfere on its second appearance, more particularly if the bleeding be in any quantity. Not to interfere unless the vessel is bleeding, must not always be understood too literally, or we will often be prevented from performing the operation till our patient is beyond our help. The hemorrhage recurs over and over again, and the surgeon, though as near as is practicable, arrives only in time to see the bed drenched, and the patient and attendant intensely alarmed. There is at the moment no bleeding, and he vainly hopes there will be no return ; and so on goes the game between ebbing life and menacing death, the loss not great at each time, but mighty in its sum, till all assistance is useless. Many a valuable life has thus been lost which might have been saved by a more decided course of action.

Few cases are more embarrassing than these to the surgeon, or require more determination and well-considered resolution to conduct to a successful issue. One is averse to act when the immediate necessity has passed ; and unless   . we be guided in our course by a knowledge of general results, more than by the immediate case in hand, we will lose many a patient. These cases form an exception to the rational surgery of the day, which prescribes inaction, unless there be immediate call for interference. There can be little doubt that hemorrhage may often be definitely arrested by pressure applied with care along an extensive part of the wounded artery, as well as to the apertures ; but such treatment is not adapted for gunshot wounds, from the depth and narrowness of their tracks, unless we so enlarge them as to admit the compress deep into the wound. This was shown in the case recorded above, as well as in many others. The discharge is pent up by the plug, and **burrows** largely among the tissues.

There were many cases of hemorrhage from the hand, succeeding gunshot wounds, which came under my notice during

the war.   Many of the injuries resulted from the accidental
explosion of the patient's own gun, and, I suspect, in not a
few cases from intention.   Hemorrhage, in such instances,
was at times very troublesome, especially when the bones of
the hand were much fractured, as it was then difficult, if not
impossible, to secure the vessel in the wound.   The secondary
bleeding usually appeared early in these cases, and, so far as
my observation went, ligature of the brachial seems better
practice, when local means fail, than putting a thread on the
vessels of the forearm, as I saw done several times in the
East.   In recent cases we can often ligature the bleeding
vessel, but in the sloughing stage, with a deep wound, and
the bones much injured, it is impossible to secure it.   To
ligature the radial and ulnar, separately or conjointly, exposes the patient to operative dangers which bring no adequate return, as the probability of success is very small.   In
the following case, the ligature of the vessels of the forearm
succeeded ; but in four other cases, in which I knew it tried
for wounds of the palm, it failed utterly.   The position of
the wound in this case made it more likely that the proceeding followed should succeed : A soldier, resting his right
hand on his musket, was struck by a ball on the web between the thumb and forefinger.   The wound seemed trivial,
but the whole hand swelled exceedingly.   On the fourteenth
day arterial hemorrhage occurred, and pressure was applied.
The bleeding repeatedly recurred, and still pressure was per-
severed in.   Finally, the radial, and then the ulnar, were
ligatured before the hemorrhage was commanded.   An early
search in the wound would probably have succeeded in
securing the vessel.

I have seen the method of pressure on the brachial **by**
flexing the arm and by bandaging ; both fail in some of
these cases.

Hemorrhage from the face **of** stumps **is** unquestionably
one of the most disagreeable complications which can arise
during their treatment.   The scorbutic state of the blood of

most of our men made their stumps highly irritable, and
liable to sanguinolent oozing. Their strength was thus
much wasted, and other complications of hardly less serious
importance were superinduced. All noticed the prevalence
of these bleedings when the hot sirocco blew, or previous to
those violent thunder-storms which did so much to clear the
air. The patients often complained at these times of feeling
"as if their stumps would burst," and the bleeding seemed
to give them much relief. The blood which flowed was
commonly more venous than arterial, thin, watery, and of a
brick-dust color. When cold air or water, combined with
elevation, failed to check it, pressure and the perchlorate of
iron generally succeeded. Its appearance was always an
indication for more fresh air, tonics, and better food.*

Guthrie counsels us, in the event of hemorrhage from a
thigh stump which cannot be commanded by the application
of a ligature to the bleeding point, to tie the main vessel, in
the first instance, at a point the nearest to the end of the
stump, at which pressure commands it, provided it be beyond the sphere of the inflammation ; and if this fail, then
to reamputate the limb. He adds, that if pressure above the
going off of the profunda is necessary to command the bleeding, then we should amputate, in place of tying the vessel in
the groin. The dictates of so great a master are not lightly
to be controverted, but, so far as my comparatively very
limited observation goes, I would be disposed to tie the iliac,
rather than either ligature the femoral high up, or reamputate the limb; this is, of course, always providing that the

---

* Briot (Hist. de la Chir. Milit.) remarks that strong, vigorous subjects are not those in whom he has seen hemorrhage, either primary
or secondary, most commonly follow gunshot wounds; but, on the
contrary, it was more common in patients of an opposite character.
This he ascribes to the want of tone in these men, preventing the
contraction or closure of the vessels. The same thing, he says,
exists in the power we have of arresting bleeding in primary and
secondary operations—those necessitated by accident and disease.

vessel could not be secured on the face of the stump. Bleeding from a thigh stump is so apt to proceed from one of the deep vessels, and to be temporarily arrested by a tourniquet applied to the femoral, but whose strap encircles the limb, that no ligature of the femoral much above the extremity of the stump could give any security against a return. The fear of gangrene in depressed subjects when the iliac is tied, is the chief objection to the practice I allude to.

The cases in which attempts were made in the East to arrest hemorrhage from stumps, by applying a ligature to the main vessel above the extremity of the stump, were, I believe, singularly unfortunate. Well-applied pressure along the course of the principal vessel, adapted to diminish the circulation through it, has sufficed, in some few most threatening cases, finally to arrest bleedings which had recurred frequently; but in these cases the implication of the main vessel was clearly made out. Take the following case as an example. The state of the vessel, as discovered **after death,** also lends an interest to the narrative. Hemorrhage took place to a slight extent, from a thigh stump, on **the ninth** day after operation, and was repeated on the following morning. A tourniquet was applied over the course of the femoral, so as to moderate the flow of blood through it. On the fourteenth day the bleeding returned, and the tourniquet was tightened for four hours, so as almost to arrest the current of blood in the great vessel, and afterward, though loosened, was still left so tight as to restrain the free flow of the blood through the main artery. On the sixteenth day the bleeding returned, and the same treatment was followed, the position of the compressing force being carefully shifted from time to time. From this period the hemorrhage never reappeared. The patient subsequently died of pyæmia, when it was found that an abscess had **formed around** the great vessels, extending **from** the end **of the stump** upward for **some** inches; that **the** artery **was fairly** opened by ulceration, to the extent of an inch **from** its termination, but beyond that distance a dense clot occupied its caliber for an inch and a quarter. The vein contained much pus. Purulent matter was freely deposited in the lungs. Here the ulceration of the end of the artery allowed the bleeding to take place, while the subsequent formation of a clot above arrested it.

The exact number of cases in which **tetanus** has followed wounds during the war, or the nature of the injuries giving rise to it, I have failed to learn from the army returns. It was not, however, by any means common. I know of six cases only which occurred in camp, and seven which took place at Scutari. The usual proportion to wounds is, according to Alcock, one in seventy-nine. We have certainly not had that ratio.* In no case, the particulars of which I could learn, did it occur after the twenty-second day—the limit as defined by Sir James M'Gregor. In the cases of which I have known the details, there was no confirmation of Baron Larrey's theory as to the set of muscles affected according to the position of the wound. The cases occurring in our army have been, so far as I know, with one exception, universally fatal. Of the six instances which appeared in front, one followed a compound fracture of the thigh, one a face wound with destruction of the eye, one an amputation at the shoulder, one a flesh wound of the leg, and the other two cases following wounds, without fracture of the thigh, unfortunately happened under my own charge in neighboring wards in the general hospital, and within a very short time of one another. Of those which appeared at Scutari, one followed an amputation of the hand, two succeeded compound fractures of the thigh, one was a frost-bite of the toes, one was a compound fracture of the leg, and of the other two cases I could not learn the primary

* In the Sleswick-Holstein war, Stromeyer had six cases in a list of 2000 wounded.

lesion.*   I give the particulars of my own cases, from their presenting some points of interest, of which not the least was their extreme similarity to one another.

---

* In the records of the army in India I have been able to find details of nineteen cases of tetanus, of whom only one recovered. The patient had received a severe burn from an explosion of powder, and was treated by the "injection and inhalation of sulphuric ether." In three cases it followed amputation ; in three, balls lodged in bone; in four, flesh wounds ; one, a penetrating wound of the chest ; one, a contusion of the face ; one, a wound of the hand ; one, a needle broken in the heel ; one, the exposure of a suppurating wound to cold air ; one, an injury of the foot ; in one, a compound fracture ; and in one, an injury of the ankle.

Alcock reports seventeen cases. In ten the disease followed flesh wounds, (three of the upper extremity, three of the lower, and one of the trunk.) In two it followed wounds of the foot ; in four, compound fracture ; in one, a primary amputation. All died except two, one of whom was treated by opiates alone, the other by opiates succeeded by large doses of carbonate of iron.

Larrey makes reference to many cases in his memoir in the campaigns in Egypt and Germany. He says it was less acute in the latter than in the former country. In Egypt I find reference made to upwards of thirty cases, the details of many of which are not given. All seem to have died within a week of the appearance of the symptoms. One followed a slight wound of the face ; another, a wound of the hand ; another, a wound of the ear. Three were cases of flesh wounds. One was an amputation of the foot ; one, an amputation of the arm ; one, a wound of the foot ; and one was caused by a fish-bone sticking in the throat. In the German campaign no numbers are given ; but of those mentioned, two were amputations of the thigh ; one, a wound by round shot, of the back : another, a similar wound of the leg ; another, a wound of the hand ; and the sixth was a lance wound of the forehead. Several of these recovered **after** section of the nerve.

In the Hôtel-Dieu, in July, 1830, they had but one case of tetanus, among the 390 patients wounded by gunshot, treated. It occurred in a perforating wound of the thorax, and was fatal. On looking over these cases, it would not appear that wounds of either the foot, or yet of the lower extremity, show a greater tendency to cause tetanus than others.

Hughes, a private of the 44th Regiment, was admitted
into the general hospital on the 18th of June. In the
assault on the Redan, a ball had entered an inch below the
anterior superior spinous process of the right ilium, and,
passing downward and outward, lodged deeply among the
muscles of the thigh. After a most careful and prolonged
examination, its position could not be ascertained, and it
was left, in the hope that in a day or two it might become
defined, or that it might perhaps remain altogether without
doing harm. There was no fracture, and no pain. The
case was treated as a flesh wound. On the thirteenth day,
the patient for the first time complained of pain behind the
great trochanter of the right side, and the presence there of
deep fluctuation caused me to make an incision. A considerable accumulation of pus was found, and in the sack of
this collection I discovered the ball much flattened. I
freely enlarged the wound, so that all retention of matter
was prevented. Next day some cloth was discharged from
this opening. On the seventeenth day his manner was
changed. He was irritable, and complained of his wound.
Pus continued to flow freely, and his general health was
unimpaired. He said that he had caught cold, and "that it
had taken him in the jaws," which were a little stiff. His
bowels being costive, I ordered him a purge, and an embro-
cation for his jaws. I had not at this time any suspicion of
the impending evil. Next day I found that his bowels had
been fully moved, **and** that most offensive dark-colored
stools had resulted. The trismus was now very marked.
The masseters were hard and contracted, like clamps of
iron. I examined the wound, and further enlarged it. A
large emollient cataplasm was applied, and a drop of croton
oil given internally. His bowels were freed **of** much more
of the same fetid dejection which he **had** voided formerly.
In the afternoon, when rising to go to stool, (which he in-
sisted on doing,) he had a violent spasm over the right side
of the **body, not** accompanied with any pain. From this

time the spasmodic contractions set in, recurring at certain
intervals, leaving him at times for half a day, but always
returning till his death. These spasms were nearly confined
to the wounded side, and affected the muscles of the thigh
most. I began the use of the acetate of morphia in gr. ij
doses, and afterward diminished it to one grain, administered every hour till he slept. This he did in snatches
during the succeeding days, waking up startled if any one
walked near his bed. Whenever he slept for an hour or two,
his symptoms were markedly alleviated. When he slept,
the opium was intermitted for some hours, and then resumed.
Only on the first day did he exhibit the slightest symptoms
of narcotism, and so much relief did he experience from the
use of the drug that he earnestly asked for it whenever he
was a few hours without sleep. He always denied suffering
any pain, though from the way in which the corners of his
mouth were drawn upward and backward, so as to expose
his teeth, and the manner in which his brow was knit, **he**
looked as if he was in extreme agony. There was one
small spot, presenting no peculiarity to the eye or touch, on
the inner side of the knee, and another on the ankle of the
wounded limb, which he always said gave him much pain.
The least pressure on these spots always caused the most
violent spasm. He frequently expressed his astonishment
at the limb starting in the way it did, and tried in vain to
prevent it. His mind remained unaffected till near his
death, when he became dull and heavy like a drunken man.
The muscles of the neck, the long muscles of the back, as
well as the *serratus magnus* of the right side, and the
muscles of the thigh and leg of the same side, became hard
as a board, particularly during the transit of **a spasm. So**
hard and contracted were they that when we had occasion
to move him in bed, he could be raised like a **log of** wood,
at least so far as his right side was concerned. His abdomen was much distended, and its muscles hard. The
clysters which were administered during the course of his

13

treatment always gave him much relief from the feeling of
" bursting " of which he so often complained. He lay diagonally in the bed, his wounded limb stretched straight out,
and the other drawn up. Latterly he suffered from a severe
pain, which continued to shoot from the ensiform cartilage
to his spine, and also from intermitting pains in his right
side. For a couple of days there was a diminution of the
discharge from the wound, but ultimately it became quite re-
established. His skin was always bathed in perspiration,
the excretion having a most pungent and offensive smell.
For some days before death a miliary eruption showed itself
over the upper part of his body. His pulse was slightly
elevated during the course of the malady, but it never
reached any very high standard. His respirations varied
from twenty-six to twenty-eight per minute. A very viscid
spittle, which he was always trying to hawk up, gave him
much annoyance. He had retention of urine, and latterly
suppression. There was some blood mixed with his urine
for a couple of days.

I have already alluded to the treatment which was followed. Purgatives, opium given freely, at first combined
with camphor, and latterly alone. He frequently took as
much as fifteen grains of opium before he slept, and altogether he must have consumed a great quantity of that
drug. He asked for fomentations to be applied to his limb,
which to the hand felt colder than its fellow. Their application, he said, gave him relief. His ability to swallow
semi-fluid food enabled me to give him the most nutritious
diet I could devise, along with wine. With intermissions
and exacerbations his fatal malady progressed. A spasmodic cough was added to his other ailments. On the
afternoon of the tenth day of attack, his symptoms greatly
abated for some hours, and while he was conversing with a
comrade, he was seized with what the orderly termed "a fit,"
vomited some dark matter, was severely convulsed, so that
the body was drawn backward and to the right side, and

before I reached his bedside he was dead. By a mistake
of the hospital sergeant, no post-mortem examination was
got. I looked hurriedly at the wound shortly after death.
The fascia lata was much lacerated, and the parts beneath
were sloughy.

Barker, a private in the 38th Regiment, aged 20, was
admitted into the general hospital in camp on the same day
as the last patient, June 18th. A ball had penetrated his
left thigh at its inner and lower aspect, and lodged. It
could not be found, though every means were used. Four
days afterward it was felt near the wound, and removed.
By the 28th the wound was looking sloughy, and the discharge was thin and unhealthy. He complained much more
than was usual about his wound, and appeared very anxious.
On the 30th I noticed some twitching of the limb as it was
being dressed. His bowels were free, but he complained of
sleeping little at night. The wound was freely enlarged,
and covered with a poultice. He was purged with croton
oil and clysters. He grew gradually worse. During the
two succeeding days, the spasms were very decidedly pronounced over the left side. He described them himself as
proceeding in "flashes" from his wound to the spine, and
back again. Touching the limb, and particularly the sole
of the foot, immediately aroused the most violent spasmodic
contractions. His pulse rose to 92, and his respirations to
29 per minute. He did not complain of pain, but was
greatly distressed by a thick spit which clung to his teeth,
and which he was always making violent attempts to expel.
The left side of the body was almost alone affected, and the
spasms, as in the last case, drew him diagonally backward,
and to the wounded side. He had no trismus for the first
day, but afterward it became marked. He always said that
he was sure, if he could only sleep, he would be all right. I
brought him under the influence of chloroform, and while its
effects continued, the spasms were relieved, and certainly the
pulse and respirations were reduced in frequency; but so

soon as he awoke, all his worst symptoms returned in undiminished vigor. Having seen the utter futility of chloroform to relieve the spasms permanently, or to arrest the disease, in two former cases at home, where the anesthetic had been fairly tried; and having many wounded to attend to, and no assistant to whom I could intrust the exhibition of the anesthetic,—I determined to abandon it and trust to opium. This, with the enemas, nourishing food, and local emollient applications, comprehended all the treatment. The symptoms were not abated, except for short intervals, and then only in proportion as sleep was procured. His skin was always covered with an odorous perspiration. The abdomen got distended and hard. The muscles of the back were markedly hard and contracted, particularly on the left side. The left leg was stretched out spasmodically, every muscle defined. The right limb was drawn up, and he lay across the bed. The wound was sloughy, and shreds of fascia escaped with the discharge. The urine became scanty and high colored, and required to be drawn off by the catheter. Eventually he suffered much pain in the left groin and calf of the leg, as well as at the ensiform cartilage. When trying to raise himself on his elbow on the fifth day of the attack, and seventeenth after admission, he was violently convulsed, so that he was bent greatly backward; he put his hand to his throat as if choking, and fell back dead.

The wound was found to be lined with an ashy slough. The bone was not injured. The fascia lata was much torn, and was pierced and ulcerated at a spot on the anterior and external aspect of the limb, some little distance from the wound. The ball had evidently penetrated to this point. No nerve fibers could be detected near the wound. The parts in the neighborhood were sound. The brain and internal organs were healthy. The lungs were only slightly congested, and viscid mucus was present in the larger tubes. The spinal canal contained a good deal of fluid blood. The cord and its membranes were congested. In the lower cer-

vical and upper dorsal region the substance of the cord was
varicose—contracted and expanded into a series of knots.
There was no other pathological appearance.

On looking at these two cases in connection, the curious
parallelism must strike one. The very distinct manner, too,
in which so many of the peculiar symptoms of this deadly
disease were developed, particularly in the first case, was
interesting. Whether opium, which appeared to act so
beneficially, had it been pushed further, so as to produce a
more decided impression, would have done good, is a question. I believe it would have affected the result but little.
The similarity between the wounds in these two cases, the
non-discovery of the balls for some days, the symptoms, the
season of the year, and the state of the cord in the last case,
were all interesting. That the high temperature we had at
that period had much to do with the production of the disease, is not certain; yet three of the six cases which occurred
in front appeared during a period of extreme heat. In one
case of tetanus, succeeding an injury of the foot, which recovered, chloroform was repeatedly administered for prolonged periods, and anodynes applied to the spine. The
particulars of this case are, I understand, to be published by
the surgeon in charge, Dr. Ward of the 17th Regiment.

As to treatment we are yet unfortunately in the dark.
Romberg sums up his review of the question thus: "The
results of treatment amount to this, that wherever tetanus
puts on the acute form, no curative proceeding will avail;
while in the milder and more tardy form, the most various
remedies have been followed by a cure." Larrey trusted
most to opium and camphor, with section of the nerve in
cases adapted for it. On reading the many accounts which
have been given of cases of this disease, opium **and chloro-**
form appear decidedly to have the greatest evidence in their
favor.

The unpublished records of the Indian campaigns illustrate to a great extent the remarkable effect which unex-

13*

tracted balls seem to exercise on the development of this
fatal affection, more especially when they lay under strong
fasciæ, as in my cases. In India, as well as in the continental campaigns, amputation at the shoulder appears to be
one of the operations most frequently followed by tetanus.

Sudden vicissitudes of temperature have been always
looked upon as most powerful causes of tetanus, especially
the change from a hot day to a cold and damp night, which
is so common in the tropics. So it was after the battles of
Ferozepore and Chillianwallah, when the wounded lay exposed to **very** cold nights succeeding days of hard work
under a burning sun. Larrey notices the same circumstances as having predisposed to the disease in Egypt, and
in the German campaign of 1809. After Bautzen the exposure to a very cold night produced over a hundred cases,
and after the battle of Dresden, when the wounded were
placed in like circumstances, they lost a very large number
from tetanus. Baudens gives a very interesting recital from
his African experience, which shows the influence of cold
and moisture in producing this disease. Forty slightly
wounded men were placed, in the month of December, and
during the prevalence of a northeast wind, in a gallery on
the ground floor, which was open to the north. Fifteen different cases of tetanus appeared in a short 'time — among
this number twelve died. The remainder were removed to
a more sheltered place, and there were no more attacked.
The exposure after the Alma might have been expected to
produce many cases; but I do not believe that many resulted therefrom, though the confusion which existed, with
regard to reports, at that period, makes it difficult to know
what was the real effect of such exposure in reference to this
point.

Opposite extremes of temperature appear to cause simi-
lar effects in this most curious affection. In both the Indies,
heat is looked upon as a most powerful predisposing and
exciting cause, and idiopathic tetanus is there not uncommon

both among the natives and the European troops, while in the arctic regions it is even more frequent and fatal. Sir Gilbert Blane tells us, that out of 810 wounded men who came under his observation in the West Indies in 1782, thirty were seized with tetanus and seventeen died. Dr. Kane's experience in the arctic regions shows how apt **exposure** to a low temperature is to cause it. He tells us that while most of his party were more or less affected, he lost two men from "an anomalous spasmodic affection allied to tetanus," and that all his dogs perished from a like cause. The great cold, exposure, and frost-bites, which were sustained in the Crimea during the first winter, were followed by fewer cases of tetanus than we might have expected, though I suspect more cases appeared than we have any record of.

I am ignorant of the total number of cases which have occurred in the French hospitals; but of five cases with the history of which I was familiar, and which appeared about the same period in the hospitals at Constantinople, one followed compound fracture of the thigh; two, wounds of the foot; and one, a penetrating wound of the chest in a Zouave, who, after recovery, was allowed to visit the city, where he remained drunk for three days: he was seized with tetanus on his return, and died in 48 hours. The French trust mostly to opium in the treatment, and report favorable results from its use; though I suspect, from what I have heard, that not a few cases of simple trismus were inadvertently classed by them under the more formidable disease of tetanus.

I have been put in possession of the particulars of some cases which occurred in India, where amputation was had recourse to in tetanus. They all ended fatally without relief, though if performed early, before the peripheral irritation had time to set up much centric disturbance, this step would certainly **appear to** promise **good** results, in so far as **the cause being removed,** local applications to the spine would have a better chance of succeeding in allaying the excited action.*

Hospital gangrene was not common in the East. During the first winter it prevailed a good deal in a mild form at Scutari, but it never became either general or severe. It did not appear to pass from bed to bed, but rose sporadically over the hospitals. It frequently attacked the openings both of entrance and exit, but occasionally seized on one only, showing apparently a predilection for the wound of exit.† At times it showed itself only in part of a wound, and spread in one direction alone. It was never severe, and was invariably, as far as I saw, of the variety designated "ulcerous" by Delpech, and "phagedæna gangrenosa" by Boggie. In many cases the best designation for it, as it appeared with us, would have been the old one of "putrid degeneration." The earliest symptom was pain in the part, which sometimes preceded the ulcerative process by a couple of days. The edges of the wound did not swell up, but remained thin as they were undermined. The pain generally continued during the process of destruction. It appeared chiefly in the lower extremities, and in wounds whose progress toward cure had been for some time stationary. It seldom burrowed far into the intermuscular tissue, but confined its ravages to the surface and the circumference of the wound. I never saw any marked gastric disturbance attend it. If it attacked the wounds of those already laboring under fever, it appeared to aggravate the fever.

The abominable state in which the barrack hospital at Scutari was during its early occupation may well have

---

* Larrey amputated successfully in several cases, and speaks highly of it in those instances which are adapted for it; but after Toulouse it failed completely in our army, though tried extensively.

† Dr. Taylor thought in India that gangrene more commonly appeared in the wounds occasioned by grape and canister. I cannot say I observed this confirmed in the East, although it seems very probable it should be so.

caused an outbreak of hospital gangrene among the broken-
down men who lay so thickly around the doors of the offensive latrines; but I cannot say that I noticed any greater
tendency to its appearance at these places than in any other
portion of the hospital. The corridors presented, I think,
the greatest number of cases. Whenever it appeared, the
patients were isolated, and sent into wards set apart for the
treatment of the disease.*

Nitric acid, applied locally, and the exhibition of the
tincture of the muriate of iron internally, in half-drachm
doses, three times daily, proved to be the most efficacious
means of stopping it as it appeared in our hospitals. The
local nature of the complaint was universally recognized,
and local measures relied on for its relief. The application
of the escharotic not only to the edges of the sore, but also
to the healthy tissues, at a little distance round the margin,
secured by far the best means of employing the remedy. A
barrier of lymph appeared to be thus thrown up around,
which prevented the spread of the peculiar inflammatory or
destructive action in the skin and cellular tissue to which it
was always confined. The attendant fever was uncertain in
its development; sometimes it preceded, sometimes it accompanied, and sometimes it followed the local outbreak. Often
there was little if any constitutional disturbance, and occasionally the fever was of a low typhoid type. The most
generous diet was always necessary, for though it may be
true, as was the case in the Peninsula, that an antiphlogistic

---

* How far this segregation into separate wards is a good plan, I
am disposed to doubt. That the malignancy of the disease is thereby increased, and the danger to the other inmates of the **hospital**
enhanced, has been the opinion generally held on the adoption of
such measures. If each patient was taken outside **of the** hospital,
and placed in a tent by himself, it would be **the most** successful way
of treating such cases. **In an outbreak of this kind,** wooden huts
would be found most excellent hospitals, **as** I know from experience
elsewhere.

treatment is at times necessary, it can be so only in strong, healthy men, who derived the disease from infection. With us the depression of the powers of life was so marked, and appeared to exercise so strong an influence, as predisposing to its outbreak, that, in place of lowering remedies, the most strengthening, including stimulants, and, above all, fresh air, were absolutely required, and were alone of any use.

Those who had suffered in camp from diarrhœa, and whose strength had thus been much reduced, more especially those whose constitutions were strongly impregnated with scurvy, were most liable to be attacked; and, in all our cases, so far as I saw, the development of the disease resulted from a lowered general health more than from specific causes. It was, in many cases, a veritable "child of the typhus." The peculiar dark hue of the face, spoken of by writers, was not common, though it was occasionally seen; but the disagreeable smell and the rounded shape of the sore were almost always present. The introduction of disinfectants into our military hospitals has done much to prevent the prevalence of this disease, which committed such ravages during the Peninsular war.[*]

The French suffered most dreadfully from hospital gangrene in its worst form. The system they pursued, of removing their wounded and operated cases from the camp to Constantinople at a very early date, the pernicious character of the transit, the crowding of their ships and hospitals, all tended to produce the disease, and to render it fatal when produced. Many of their cases commenced in camp, but the majority arose in the hospitals on the Bosphorus, where the disease raged rampant. In the hospitals of the south of France it also prevailed, and, from what M. Lallour, surgeon to the "Euphrate" transport, tells us in his paper on the subject, it must have committed great

---

[*] See paper by Staff-Surgeon Boggie, in the first part of the third volume of the Transactions of the Medico-Chirurgical Society of Edinburgh.

ravages in their ships, from one of which, he says, sixty
bodies were thrown over during the short passage of thirty-
eight hours to the Bosphorus. With them the disease was
the true "contagious gangrene," and attacked not only
open wounds, but cicatrixes, and almost every stump in their
hospitals. They employed the actual cautery, after the
manner of Delpech and Pouteau, with apparent success, to
arrest it. The perchlorate of iron, charcoal, the tincture of
iodine, lemon-juice, etc , they employed as adjuvants. In
both the French and Russian hospitals, gangrene was often
combined with typhus, and in such cases the mortality was
fearful.

In the Crimea, during the heat of the summer of 1855,
after the taking of the Quarries and the assault on the great
Redan in June, not a few amputations of the thigh were
lost, from moist gangrene of a most rapid and fatal form.
In the case of a few who lived long enough for the full
development of the disease, gangrene in its most marked
features became established; but most of the men expired
previous to any sphacelus of the part—overwhelmed by the
violent poison which seemed to pervade and destroy the
whole economy. This form of the disease occurred in four
cases under my own charge, in men who had had a limb
utterly destroyed by round shot or grape. In all the knee-
joints were crushed, the collapse was deep and prolonged,
and the operation performed primarily in the middle third
of the thigh. Three of the four were of very intemperate
habits. All these cases took place about the same time, at
midsummer, when many other similar cases appeared in
camp. The wards, though full, were not overcrowded, and
could, from their construction, be freely ventilated. The
weather was sultry, and cholera was in the camp. The
atmosphere was surcharged with electricity, and the dreaded
sirocco prevailed. Wounds generally assumed an unhealthy
aspect for days when this pestilential wind blew. The cases
of all those who died in my wards seemed to be doing per-

fectly well up to sixteen hours, at the furthest, before death.
Three of them were seized on the eighth day after amputation, just as suppuration was being established. The
fourth died on the fifth day. The seizure and consequent
symptoms were identical in them all. In recording one
case I relate all. During the night previous to death,
the patient was restless, but did not complain of any particular uneasiness. At the morning visit, the expression
seemed unaccountably anxious, and the pulse was slightly
raised. The skin was moist, and the tongue clean. By
this time the stump felt, as the patient expressed it, heavy
like lead, and a burning, stinging pain had begun to shoot
through it. On removing the dressings, the stump was
found slightly swollen and hard, and the discharge had
become thin, gleety, colored with blood, and having masses
of matter like gruel occasionally mixed with it. A few
hours afterward, the limb would be greatly swollen, the
skin tense and white, and marked along its surface by prominent blue veins. The cut edges of the stump looked like
pork. Acute pain was felt. The constitution, by this time,
had begun to sympathize. A cold sweat covered the body,
the stomach was irritable, and the pulse weak and frequent.
The respiration became short and hurried, giving evidence
of the great oppression of which the patient so much complained. The heart's action gradually and surely got
weaker, till, from fourteen to sixteen hours from the first bad
symptom, death relieved his sufferings. All local and constitutional remedies which could be thought of were equally
powerless: nothing could relieve the system from the weight
which seemed to crush it, or enable it to support the severe
burden. Strong stimulants were the only remedies which
appeared to retard the issue for a moment. Post-mortem
examination, instituted shortly after death, showed the
tissues of the limbs, and in many cases those of the internal
organs also, to be filled with gas, and loaded with serous
fluid. The vessels leading from the stump were healthy,
and in only one case had there been any actual mortification

previous to death. The intestines, in two of the four cases, were much diseased. Was the cause which gave rise to this affection referable to "weakness or defective powers of action," arising from the patients' bad state of general health, or "excessive irritability or disposition to act," from their being of intemperate habits? or was it "excessive irritation or excitement to act," arising from the severity of the injury sustained? After the taking of the city, in September, the same form of disease again appeared, especially among the Russians who had been operated on; and was so deadly that in no case, which I could hear of, did recovery follow.*

Erysipelas was latterly rarely seen in our hospitals. Several cases which appeared in my own wards readily yielded to treatment. At Scutari there were a good many cases, at the time when the men were most depressed by their hardships; but it was seldom virulent.

The troops suffered greatly, during the first winter which they passed in the Crimea, from **frost-bite.** Death not unfrequently followed on the injuries it occasioned. The severity which marked these lesions did not arise from the degree of cold, as the temperature was never so low as of itself to cause the severe results produced, but rather from the depressed vital power of the soldiers, who could not resist the effects of a degree of cold which would have little injured them if they had been in rude health. The practice, which was nearly universal, of sleeping in their wet boots, aided greatly in causing the results. This custom arose from the fear that, if the boots were put off, they could not be drawn on again. They were retained, and thus the feet,

---

* Dr. Taylor, in his report on the 29th Regiment, to which interesting document reference has been already made, says: "It is to be observed, as illustrating the possibility of gangrene infection lying dormant for some days, or of fomites of the disease hanging about the clothing of the men, that wounded men discharged fit to rejoin their regiments were in several instances returned from camp to hospital with hospital gangrene."

kept for a long time at a low temperature, with the circulation retarded, at length lost their vitality—slowly, but all
the more surely on that account.

The scorbutic poison, too, with which the men were
drenched, predisposed strongly to the action of the **cold**,
and it was even at times difficult to say how much of the
destructive result was due to the one cause or the other.
During the first winter the frost-bites were much more severe
than during the second, and much more difficult to manage,
from the more depressed vitality of the patients. I referred
in a previous chapter to that peculiar effect caused in the
feet by the union of scurvy and frost-bite, to which it is so
difficult to give a name.

Tetanic symptoms resulted in a few cases from frost-bite
injuries of the feet. The French suffered more than we did.
In their hospitals a limb might be seen sphacelated half
way to the knee. Uncontrollable diarrhœa was a common
complication in such cases, and invariably, according to M.
Legouest, caused death. Scrive (Mem. de Med. et de Chir.
Milit., vol. xvii.) tells us that on the 21st January, 1855,
with the thermometer at 5°, they had 2500 cases of **frost-
bite** admitted into their ambulances, of whom 800 died, and
that at that period no operation succeeded, so that "it was
necessary to abstain from operating." M. Legouest says he
found, in treating his cases at Constantinople, that a solution
of sulphate of iron formed the best dressing, but of its use
I had no experience. To obtain the separation of the scars,
and regulate the subsequent granulation on general principles, was what had chiefly to be attended to. Soothing
applications appeared to be the best in the cases which I
had an opportunity of watching.

It is not easy to decide whether or not we should operate
in such severe cases as sometimes occur, when half the foot,
for example, or the lower part of the leg, is implicated.
Either step is somewhat hopeless; but if the part be unquestionably dead, and of such a nature and size as not to
be separable by "the parsimonious industry of nature,"

without so long a period of irritation and suppuration as **will be,** in all probability, fatal; or, if the presence of a **large** gangrenous surface endangers not only the patient himself, but also his neighbors in the ward; or if there is hospital gangrene present in the hospital, to an attack of which the long open state of his wound will so much expose him, then it is a fair question to consider, whether amputation is not a lesser evil than waiting. The success which follows will depend much on the state of the patient's general health, and on the condition of the parts; as, unless a clear line of separation be formed, and the parts above be tolerably healthy, the irritation occasioned by removal will be sure to cause gangrene in the stump—at least so it was in the Eastern hospitals, in all the cases in which I knew it tried. I never heard of any amputation performed under the above circumstances succeed during the first winter, but several such occurred during the second. Operating at some distance beyond the spread of the disease was generally found safer than at the place of division between **the** dead and living parts.

Any wounds from frost-bite are peculiarly difficult to heal. Many suffered from their effects for **months** after getting to France or England.

The removal of **bone** from the **toes** or fingers, however black **and** apparently dead, and though only attached by the most slender connection, was certain to cause a great amount of irritation, which sometimes became most alarming. This result was probably as much due to the enfeebled state of the patient as to the cause for which the operation was performed. Complete non-interference during every stage of treatment, the use of the mildest dressings, the removal **of** parts only when quite disjoined, proved the best line of procedure. I never saw any other followed, **either in** our hospitals or in those of the French, **without** there being ample cause to regret it. **Any** roughness even, in dressing these **injuries,** endangered the appearance of gangrene, on the verge of which they always seemed **to hover.**

CHAPTER VII.

INJURIES OF THE HEAD.

From April 1, 1855, to the end of the war,[*] the returns show a total of 630 cases of gunshot wounds of the head attended by contusion merely, more or less severe, and 8 deaths are recorded among these cases. Of gunshot fracture without *known* depression, 61 cases appear, and 23 deaths therefrom. Of cases of fracture and depression, followed by sensorial disturbance, 74 cases are mentioned, and 53 deaths therefrom; while of wounds penetrating the cranium, 67 cases and 67 deaths are recorded. Of 19 cases in which the skull was perforated, all died. The trephine was employed 28 times, and of these cases 24 ended fatally.[†]

MR. GUTHRIE has said, with much truth, that "injuries of the head affecting the brain are difficult of distinction, doubtful in character, treacherous in their course, and, for the most part, fatal in their results." Of all the accidents met with in field practice, these are, beyond doubt, the most serious, both directly and remotely—the most confusing in their manifestations, and least determined in their treatment, although they have engaged the attention of the master-minds of all ages and countries, from the time of the old surgeon of Cos down to the present day. Such men as

---

[*] The returns are not complete before the date specified.

[†] Alcock reports 28 cases of fracture of the skull from gunshot, and 22 deaths. Menière gives 10 penetrating wounds by balls, all of whom died—half on the day of admission. In the medical reports from India I find only 9 cases so detailed as to be useful. They were all penetrating wounds, and 6 of them died. Lenté, in his statistics of the New York Hospital, mentions 128 cases of fracture of the skull, attended by death in 106 instances. Several of these were fractures of the base, and none by gunshot.

Petit, Quesney, Ledran, Pott, Dease, Heister, Cooper, Dupuytren, Bell, Velpeau, Larrey, Brodie, and a host of other honored names, have thrown the light of their large experience and commanding genius on the subject; even minor points connected with it have been made the theme of whole libraries, and of innumerable discussions in the first medical societies of the world; still there is no accident which the surgeon takes charge of with more fear and hesitation, as in no class of cases does he feel so much the mystery which surrounds and guards our life: for while in some cases death follows the most trivial injury, in others a vast amount of destruction, and even removal of brain-matter, causes little if any disturbance.

In war, injuries of the head of all descriptions are presented to us. Those by *contre-coup*, especially such as implicate the **base of the skull**, are certainly rare; but these also at times do occur. The comparative rarity of this form of injury in military, as compared with civil practice, is possibly accounted for by the less frequent occurrence of **such** accidents as are fitted to injure the base, and by the **fact** that war-projectiles seldom present a surface so large as to supply those conditions which **the** experiments of Bichat would show **are necessary** to produce fracture of the skull by counter-stroke. **It is,** however, by no means true that the "punctured fracture," as it is termed, is the only species **of injury to** which soldiers in the field are liable. Shell, grape, and sword wounds of the skull afford examples of almost every **kind** of fracture.

The nature of the injury inflicted by a ball striking the skull will depend chiefly on the angle of incidence, and **the** velocity. The character of the **ball**, too, has **more to do** with the matter than is generally supposed. **If the direction** of the projectile be very oblique to **the surface**, and if the force be exhausted **at the moment of** contact, then the injury may be very slight—a mere contusion of the soft parts

14*

or of the bone.* If the force be greater, then the peri-cranium may be much injured, the bone considerably bruised, or slightly fractured throughout its whole thickness, or in one or other of its tables separately—the fracture of the inner sometimes taking place without any apparent injury of the outer.† Further, the brain may be injured as well as its case, when the blow is yet more direct or severe. This injury may be merely of such a nature as, John Bell well says, "we choose to express our ignorance of by calling it a concussion," which may pass away, doing little harm, or which may be followed, at an uncertain interval, by enceph-alic inflammation, and compression from effusion.

Again, the effect of a ball "brushing" over the skull may be such that, while the bone is not fractured, the vessels between the skull and the dura mater may be ruptured, or the longitudinal sinus may be opened, as occurred in one case in the Crimea, and which has been related by the sur-geon of the 19th Regiment, in which it was observed.‡

A remarkable instance, showing how completely the skull may be destroyed by a glancing shot, without the scalp being implicated, occurred at the Alma. A round shot, "en ricochet," struck the scale from an officer's shoulder, and merely grazed his head as it ascended. Death was instant-aneous. The scalp was found to be almost uninjured; but so completely smashed was the skull that its fragments rat-tled within the scalp as if loose in a bag. The condition of the brain was, unfortunately, not examined.

---

* Stromeyer supposes that the danger of a grazing shot arises very much from pyæmia. Inflammation of the bone follows the injury, the veins of the diploe become implicated, and thus pus enters the system.

† Preparation No. 2594 in the museum of Fort Pitt exhibits the manner in which part of the outer table of the skull may be removed by gunshot injury, and yet the inner wholly escape; and No. 2511 shows that the inner table may be "considerably depressed, without corresponding depression of the outer."

‡ See *Lancet*, vol. i., 1855.

A bullet, from the great force with which it impinges upon the skull, and the concentration of that force on a small point, causes a fracture dissimilar to most of those which are met with in civil practice. It is this concentration of the force on a small point which renders fractures from a ball so dangerous, as the bone is driven deeply into the brain, and the splintering, especially of the inner table, is often very severe.

The greater splintering of the inner than of the outer table, by a ball penetrating the skull from without, is explicable on the principles which interpret the difference between the wounds of entrance and of exit in the soft parts, and which I before explained by reference to a series of experiments on planks of wood. The greater support afforded to the outer than to the inner table, by the parts lying behind it, and the diminished force of the ball as it passes through each, sufficiently account for the difference. An observation of Erichsen's on the point quite supports this explanation. He has noticed that the characters of **the** apertures in the two tables were reversed, in an instance in which a man had committed suicide by shooting himself through the head from the mouth—the ball thus passing from within outward. In a case from Bagieu, related by Sebatier, the same circumstance is noted in a similar instance.* The preparation in the Fort Pitt museum, numbered 2592, illustrates the same thing. In that case, a ball had perforated the head, thus making two holes, the one in

---

* Larrey thinks that in young persons a ball may enter the skull, leaving a hole less than itself, from the yielding and subsequent closure of the osseous fibres. This is not observed in **the old, in** whom the bone is more brittle, and splinters. A case **is related** by Dr. Longmore, **of the** 19th Regiment, in the second volume of the *Lancet* for 1855, by which it would appear that **a** ball may split, part enter the skull, **and yet** the bone **recover its** level by its resiliency so completely as to leave **no trace of** the passage of the part of the ball which entered.

the front and the other in the lateral and posterior part.
The inner table of the orifice of exit is regular, while the
outer "is torn up to an extent much larger than the ball."
An appreciation of these distinctions is of much use to the
medical jurist.

The character of the fracture caused in the skull by the
large conical balls is, I am inclined to believe, considerably
different from that occasioned by the round ball. The destruction by them of the outer table always appeared to me much
greater than by the round ball; and thus, perhaps, it is that
the size of the openings in the two tables is more equalized
in the wounds occasioned by the former than by the latter
species of missile. So it comes, I think, that the true "punctured fracture" is less seen now in military practice than it
was formerly. I state this, however, with much hesitation,
as it would require a larger number of observations than I
possess to substantiate it.

Balls striking the head otherwise than perpendicularly to
its surface, or impinging against one of its angles, may be
split—part entering the skull and part flying off. This occurred in cases which have been related by Mr. Wall, **of the**
38th Regiment, and by Dr. Longmore, of the 19th.* Such
instances are not uncommon in war. Larrey, following the
half which entered, removed it by counter-opening from the
back part of the head. One half of a split ball has been seen
to lodge between the tables of the skull. The whole ball,
also, has been found thus placed, especially at the fore part
of the head. There are various instances on record of a
round ball penetrating the outer paries of the frontal sinus,
without injuring the inner table; but I believe that no such
instances will **ever be** found where a conical ball is **used.** *It*
not only penetrates, but generally perforates, the **skull, and**
almost always proves fatal.

The most dreadful injuries **of the head seen in war are**

---

* See Addenda to the last edition of Mr. Guthrie's *Commentaries.*

those occasioned by shell. Although rarely, yet it does at times happen that this missile cuts open the scalp only, or merely grazes the bone ; yet it more frequently occurs that large masses of the skull are driven by it into the brain. Examples will be afterward given of shell wounds of the head. One of the most ghastly injuries of the skull which I ever witnessed was caused by a fragment of shell. The whole frontal bone was driven deeply into the brain, yet, strange to say, the poor sufferer lived for twenty-four hours after such a wound.

Sword-cuts sometimes, as is well known, slice away parts of the skull. These portions will, at times, readhere, if immediately applied. In the museum of the Val de Grace several remarkable examples of this are to be seen.* I had under my charge, after the fall of Sebastopol, a Russian soldier who had received such a wound, although the bone was not entirely detached in his case. The left parietal bone was cleft so as to be almost separated. He would allow no one to touch his wound except a comrade. His **recovery** was complete, the brain never showing any tendency **to protrude**, although quite visible throughout the whole extent of the wound. I saw this Russian in the interior, after peace, in perfect health. The comparative rarity of hernia of the cerebral substance after sword, as compared with gunshot wounds, is very remarkable.

Cuts from a blunt sword are peculiarly dangerous, from the extensive splintering and depression of the inner table which so commonly results.†

---

* Sebatier relates several such cases at length, from Léaulté, Platner, and others.

† That trephining does little good in these cases, **is illustrated by** the practice of Dease, who had under his charge many men wounded by the "hanger," which **played so important** a part in all the street frays of his time. **Four of the seven cases he** trephined died; while in the only four instances in which he seems not to have interfered, recovery followed. In a case which, although not caused by

One of the most remarkable circumstances connected with gunshot wounds of the head is that they are not more universally followed by concussion, or that the symptoms of concussion, when produced, are often so temporary in duration. I have been frequently told by men who had received wounds of considerable severity that they experienced merely feelings of passing "weakness" when struck. Symptoms of concussion, however, more generally follow severe blows; and the gradual and almost insensible manner in which this state passes into one of compression or of inflammation, and that into consequitive compression, forms one of the most treacherous and dangerous features of these cases. It is

---

a sword, was yet a fracture of a similar description, and which occurred lately in the Royal Infirmary of this city, under Mr. Lyon, the recovery was probably owing to the non-interference with the injured bone, further than the removal of loose portions. A man aged 19 was admitted on the 28th of July. He had been struck on the head by the handle of a crane, and the whole scalp round and round the head, with the exception of the anterior part, separated. The bone was fractured into small pieces, to the extent of four inches by one and a half, over the right side of the head—the fracture slanting obliquely over the orbit. He suffered much from the shock when admitted, but replied to questions put to him. Bony spiculæ were driven into the right eye, the right malar bone was broken, and the frontal sinuses opened. The loose and broken bones were removed, when the brain was found to be laid bare to the extent of three inches, and the dura mater destroyed. Low diet, purging, and cold locally, were the remedies—the scalp being carefully laid down, and the spiculæ removed from the eye. The fractured bones were not interfered with, further than the removal of perfectly loose portions. A week after admission, the brain began to protrude by the opening in the skull, but by gentle compression it soon receded, and the patient made a rapid recovery, interrupted only by a slight hemorrhage from a vessel in the scalp, which was easily suppressed. The wound completely healed—the bone being bridged over by dense tissue, and the cicatrix sunk in a narrow furrow, the pulsations of the brain remaining visible. His pulse never exceeded eighty. The supra-orbital ridge remains much below its proper position, and the right eye is destroyed.

evidently a matter of much importance to those who advocate trephining, in certain circumstances, to be able to distinguish accurately between these variable conditions; as to operate in cases of mere concussion, or in a state of inflammation, would be murder, yet how to discriminate is a practical puzzle in many cases—especially in a large number which fall to be treated in the field, when the period of their coming under observation is very uncertain, and when no account can be got of their history or early symptoms.   It requires but the most cursory reading of surgical works to determine that the utmost confusion has always existed between these various pathological conditions; even Sir Astley Cooper, with all his habitual clearness, has not unfrequently confounded them.   It is little wonder that it should be so, as their clear distinction is found only in books, and their interdependence and mutual reactions, as well as the uncertainty of their respective manifestations, all contribute to deceive "the pride of our penetration," and lead us into error.

The absence of any ascertainable cause, and the threatening symptoms which were present, in the following case of concussion, interested me a good deal at the time.   In former days it would have been infallibly ascribed to the wind of a ball.   Quin, a private in the 18th Royal Irish, suddenly fell down unconscious, in the advance on the Redan, early in the morning of the 18th of June.   He never could tell how this happened, not being aware of any injury.   He was brought into my ward insensible a few hours afterward. His symptoms were those of severe concussion.   The surface of his body was cold, his respiration was slow and regular, and his pupils were contracted.   No injury could either then or afterward be discovered.   Warmth, and an enema of the arom. sp. of ammonia, helped to restore him to consciousness, after he had vomited.   **He continued**, however, for some hours, like a man half drunk.   Reaction was so violent **as** to call for bleeding, cold to the head, antimonials, and

purging, to moderate it. Some days afterward he suddenly
became delirious, with injected eyes, one pupil being contracted and the other a little dilated. He complained much
of his head, which he afterward said had felt all the time as
if strongly bound by a cord. There was never any paralysis or subsequent unconsciousness. By free purging, shaving
the head, applying cold, restricting him to very meager diet,
and, latterly, by the use of blisters to the nape of the neck,
he completely recovered, though for about a month he suffered from severe headache, double vision, and a pulse
unusually slow, and little changed on assuming the erect
posture.

The danger occasioned by gunshot wounds of the head
will depend much on the part struck. At some places the
ball is more apt to glance off than at others, while the strong
processes of bone, the situation of blood vessels, and the
apparently greater necessity to life of some parts of the brain
than others, introduce many elements into the calculation of
the result. Notwithstanding all this, however, the curious
eccentricities which characterize these injuries—the slight
disturbance created by some which, to all appearance and
experience, are ten times more severe than others that prove
fatal—upset our preconceived opinions; and, while they
puzzle us to account for the difference, they prove the truth
of Liston's aphorism, that "no injury of the head is too
slight to be despised or too severe to be despaired of."

Generally speaking, it appears tolerably certain that
wounds of the side of the head, especially **anterior to the
ear**, are the most dangerous to life; and that a descending
scale will give the following order—the fore part, the vertex,
and the upper part of the occipital region, the last being
decidedly the least dangerous. Remarkable exceptions to
this graduating scale of danger do, however, occur.

There are, at the same time, other circumstances besides
the seat and nature of the injury which influence the result.
The age of the patient is, perhaps, the most important of

these. With children and young persons, the same gravity by no means attaches to the prognosis of head injuries as to similar accidents occurring to the old. Mr. Guthrie has well observed, that in the accounts of wonderful escapes and successful operations on the head, the subjects have been, in general, below puberty. The temperament of the patient, his excitability, his social condition, as giving rise to more or less anxiety regarding the result of his case; the means there are of carrying out his treatment as to quiet, isolation, etc.; the place where he is treated, whether in the hospital of a populous city, where the results of such cases are usually so fatal, or in the country, where so much more can be accomplished,—all these are important items in forming an opinion regarding injuries of the cranium.

Gunshot wounds of the head, being chiefly **compound**, enable us to ascertain, with tolerable precision, the amount of injury which has been inflicted; and if it be thought necessary to employ any means to elevate depressed bone, we can do so with less hesitation than if the scalp were unhurt; as, if it be true, what some of our best surgeons tell us, that the danger of inflammation in the membranes is increased by opening the integuments, then this source of danger cannot be charged to us.* Such facilities should not, however, make us less careful in our proceedings.

As to the use of the **trephine**—the cases, and time for its application—less difference of opinion, I believe, exists among the experienced army surgeons than among civilians; and I think the decided tendency among them is to indorse the modern "treatment by expectancy," and to avoid operation except in rare cases. In this, I believe, they judge wisely; for, when we examine the question carefully, **we find** that there is not one single indication for having **recourse** to operation, which cannot, by the adduction of pertinent cases,

* This source of danger is particularly dwelt upon by Stromeyer, Larrey, and Dupuytren.

15

be shown to be often fallacious; while, if we turn to authorities for advice, we find that not a great name can be ranged
on one side which cannot be balanced by as illustrious on
the other.

Simple contusion, without fracture or depression, caused
the old surgeons to "set on the large crown" of a trephine
in order to prevent future danger. Fracture, although not
accompanied by depression, or any other untoward symptom,
called for the trephine in the practice of the Pott school;*
while many, even now, would operate to cure the local pain
which so often remains persistent at the place of injury.
Other surgeons, again, discarding and condemning all this,
say we should trephine only when there is depression; but
the amount of depression which demands it, each interprets
according to his own fancy. None knows so well as the
army surgeon how very considerable a depression may exist,
especially at some parts of the head, without any injury to
the brain; nor how innumerable are the cases in which great
depression has been present, without causing harm at any
subsequent period of the patient's life.

A musket-ball being the wounding cause, would appear to
some a sufficient reason why the trephine should be applied,
however slight may be the lesion. "We should always trephine," says Quesney, "in wounds of the head caused by
fire-arms, although the skull be not fractured." "All the best
practitioners," says Pott, "have always agreed in acknowl-
edging the necessity of perforating the skull in the case of a
severe stroke made on it by gunshot, upon the appearance of
any threatening symptom, even though the bone should not be
broken; and very good practice it is." Boyer and Percy are

* This most false doctrine was that also taught by the Academy of
Surgery and the leading men in France, till, by the able writings
and practice of Bichat and Desault, it was in a great measure
rejected; however, it is from the writings of M. Malgaigne, more
perhaps than from those of any other, that this question has received
its true interpretation.

equally urgent when a ball has caused the injury.   However, "the experience of war," to which Quesney appeals in confirmation of his opinion, now-a-days completely condemns the practice, whatever it may have done formerly.

Further, "symptoms of compression" setting in early or late, are laid down by others as urgently demanding the removal of the bone.   "No injury," says John Bell, "requires operation except compression of the brain, which may arise either from extravasated blood, or from depressed bone, or matter generated within the skull."   But, unfortunately, we can seldom diagnose the existence of compression with any amount of certainty, when it sets in early, and experience teaches us that each and all of those signs which are said to indicate it may, under appropriate treatment, pass away without interference; especially when these symptoms appear early, and often also when they set in late.   Compression too, when it appears at a late date, if it arise, as it generally does, from the presence of pus, is well known to be seldom relieved by trephining.   Dease first showed how it was that the matter was commonly deeply placed or diffused in such cases; and the instances in which it has been found on the surface, or evacuable by such a bold maneuver as the well-known thrust of Dupuytren, are exceedingly rare.

Some authors, again, would have us trephine only when the symptoms of compression are severe, go on increasing in severity, and have continued **for** some time; yet, **even** under such circumstances, "recovery not seldom disappoints our fears, and mortifies us by our success."*

But, finally, it is to those surgeons who instruct us to operate when certain pathological conditions exist, which **they** carefully define, but which experience, unfortunately, **tells us**

---

* See, especially, as good instances **of this, Quesney's first** and second observation**s**.   **In the first, the stupor and delirium lasted** three months, and in the other, **it had continued also** for a lengthened time.   Stromeyer, by antiphlogistic remedies alone, saved several in which the "stupor had lasted for weeks together."

do not often manifest themselves by any recognizable signs, that we are chiefly indebted for useful directions to assist us in cases of difficulty. What good can it do to say you must trephine when the internal table is splintered more extensively than the external, when effusion has taken place on the brain, and so on, when we have often no means of knowing when these conditions exist, or when we are fully aware that they have, each and all, been present, and that to a very considerable extent, without any of their appropriate signs being manifest?

But to refer more particularly to those cases which fall to the charge of the military surgeon. There are three classes to which the trephine is still occasionally applied : 1st, fracture with depression, before symptoms have appeared; 2d, fracture with depression, attended immediately with signs said to indicate compression; 3d, fracture with or without depression, followed at a late period by symptoms evidencing compression.

It is with reference to the first class of cases that "the experience of war" is most useful and most decided. There are, I believe, very few surgeons of experience in the army now-a-days who approve of "preventive trephining."* It may be said in our time to be a practice of the past—a practice to be pointed at as a milestone which we have left behind. A very large number of instances fell under my own

---

* "That blood may be effused," says Guthrie, "and matter may be formed, is indisputable, even under the most active treatment; but that any operation by the trephine will anticipate and prevent these evils, cannot be conceded in the present state of our knowledge; and the rule of practice is at present decided, that no such operation should be done until symptoms supervene distinctly announcing that compression or irritation of the brain has taken place. It is argued that, when these symptoms do occur, it will be too late to have recourse to the operation with success; this may be true, as such cases must always be very dangerous, but it does not follow, and it never has been, nor, indeed, can it be, shown that the same mischief would not have taken place if the operation had been performed early."

notice in the East, in which, by the use of evacuants and quiet, and the absence of all operative interference, a perfect and uninterrupted recovery followed these injuries, even when the bone was very extensively depressed. Every surgeon in the army can recount many such cases. *If* any patients were lost from not having been operated on, I never saw any of them; but I do know of some patients who died because they were subjected to operation.

The wonderful manner in which the **brain accommodates itself to pressure** has been remarked in all times, and the crania in our museums show how extensive the depression may be, and yet the brain escape injury, or in which, although the central mass may be pressed upon or hurt, recovery has yet followed. In the cases of fracture with **depression** which have presented themselves to me during the war, the symptoms and the amount of depression have seldom been in correspondence.* But, in order to attain

---

* Hennen, in particular, refers to a case in which bone was depressed in "a funnel shape," to the extent of an inch and a half, and yet the patient lived in comfort for thirteen years. Stromeyer mentions forty-one cases of fracture with depression from gunshot, and in many of which it is probable that the brain was injured, **although that could not be ascertained.** Of these cases only seven **died, and one** of these perished by typhus fever. All the rest recovered, and in only one case was there any operative interference, although signs of secondary compression appeared in several. The antiphlogistic treatment, carefully carried out, was alone adhered to. Seutin, who was at the head of the medical service at Antwerp when it was besieged in 1832, gives us the results of his experience in the **following words:** "Far be it from us the pretension to decide the question which divides practitioners of the greatest merit: **we will** not take up the defense of either the one side or **the other, but we** think that it is necessary to limit to a small number the cases of fracture which demand the operation of **trephining—an** operation which often causes grave accidents, **and the success** of which is always very uncertain. The **following** facts, collected at the siege of Antwerp, prove, in an evident manner, that in the greater number of cases of fracture of the skull, when they are simple, or even

15*

favorable results, it is absolutely necessary that great attention be paid to the management of the patient, of which I
shall speak more afterward.

Those who have read with attention the records of campaigns must have often been struck with the numerous instances which are there recounted where men, with gunshot
depressed fractures of the skull, have recovered in circumstances which forbade any attention being paid to them.
During hurried retreats and forced marches, this has often
occurred.   When privation was added to the absence of all
surgical interference, these happy results were the more
marked.   In Larrey, Guthrie, Ballingall, and in the Indian
reports, many illustrations of this are found.   Dease, also,
long ago recorded the observation that "those patients who
neglected all precepts, and lived as they pleased, just did as
well as those who received the utmost attention;" at which
we need **not wonder, when we remember in** what "the
utmost attention" consisted.   **Thus it would seem as if**
severe fatigue, irregular and it might be intemperate diet,
are less injurious to men with fracture of the skull than the
probings, pickings, and trephinings which form the more
orthodox and approved practice.   Deputy-Inspector Taylor,

---

comminuted, or with slight depression, we can often abstain from
operating. It was by immediate incisions, and taking care to extract
all underlying fragments, and employing mild dressings, and using
antiphlogistics and revulsives, that we have been able to avoid the
use of the trephine.   It was by such methodic treatment that we
have obtained such happy results in the case of the large number of
'wounded which have fallen under our charge."

The reunion of bone which has been depressed, with the rest of
the skull, is well illustrated by preparations 2506, 2507, and 2512,
in the museum at Fort Pitt.   In that numbered 2512, "part **of the**
squamous portion of the temporal, and part of the **parietal bone,**"
is depressed three-quarters of an inch from the original **level, and
the** diameter of the fracture is about three inches, yet the patient
recovered perfectly, and lived as an officer's servant for three years,
when he died of fever.

in his able report on the wounded of the 29th, in India, after
referring to several wonderful recoveries from gunshot depressed fracture of the skull, very appropriately remarks,
that he attributed the fortunate results in these cases "to
the system adopted of very cautious meddling with the
wound."*

---

* I cannot deny myself the pleasure of recording a case which
lately occurred in the practice of Dr. George Willis, of Baillieston,
in the neighborhood of Glasgow, which is remarkable for the extent
of the lesion, the period when the trephine was applied, and the perfect and rapid cure. William Donald, aged 36, a pit-sinker, a man
of intemperate habits, but of strong frame, was struck on the 29th
of June last, at four o'clock in the afternoon, on the left side of the
head, by a piece of stone weighing thirty pounds, which had been
thrown high into the air by the explosion of a mine he had constructed in the prosecution of his work. He immediately fell down
insensible, and was put, in that condition, into a cart, and conveyed
to his house, which lay two and a half miles from the place where
he met with the accident. In about half an hour from the moment
he was struck, and before he reached home, he slowly regained consciousness, and on his arrival at his own door he was able to walk
into the house with assistance. He was, however, unable to speak.
Dr. Willis saw him about this time, and found a semilunar wound,
about nine inches long, extending over the left side of the head, and
curving over the ear. The flap of the scalp hung down over his ear,
and a clot of blood covered the bone. On clearing away this mass
of effused blood, the bone was found to be comminuted and depressed
in an irregular crescentic shape, to the extent of four inches long by
two broad. It was driven downward to the depth of a quarter of an
inch, and comprised part of the frontal and a portion of the parietal
bones. The flap of the scalp was cleaned and replaced, and cold
applied. Nothing else was done that evening. His pupils remained
unaffected at all times, and his pulse never was much disturbed, but
at the evening visit his mouth was found drawn to the left side.
Next morning at ten o'clock the speechlessness remained, but no new
symptoms were added. The fractured bones were so firmly impacted
that they could not be removed without the use of the trephine,
which was accordingly applied at the upper part of the fracture, and
when a piece of bone was thus removed, the rest were easily got at and
withdrawn. The dura mater was entire, and rose immediately in the

More difficulty exists as to the treatment of the second class of cases referred to before, viz., those in which there is fracture with depression, attended immediately by those signs which are usually said to indicate compression.

Compression is undoubtedly the evil against which the trephine is generally employed. But yet, with all that has been said on the subject, in books and lectures, I question whether we are sufficiently acquainted with the nature, seat* or signs of compression, to warrant us in undertaking, at

---

wound. At each pulsation of the brain blood flowed from between the skull and the membrane. Whenever the depressed fragments were removed, the tongue could be protruded, which before the operation it could not. It projected to one side. The speech did not return. The scalp was replaced and fixed; he was purged and put on low diet, and kept quiet, cool, and in the dark. By night he had again lost all power over his tongue, but recovered it next morning, and from that period his convalescence went on so rapidly that in three weeks his wound had completely cicatrized; he never had an uneasy feeling, and returned in perfect health to his work within six weeks of the period when he met with the accident. I saw him, by the courtesy of Dr. Willis, some time afterward. He told me he never had had a headache since the day of his dismissal, although he acknowledged to have been repeatedly drunk. The cicatrix was firm, and considerably sunk, and the brain pulsations could be obscurely felt at one corner of the wound.

* I have myself known the trephine applied, in two cases, to injuries on the vertex of the head, when the compressing fracture existed at the base. Are we, in cases of doubt, to proceed as Heister directs? "Sometimes it is impossible," he says, "to discover the particular part of the cranium which is injured; the patient in the mean time being afflicted with the most urgent and dangerous symptoms. In these cases it will be necessary to trepan first on the right side, then on the left side of the head, afterward upon the forehead, and lastly upon the occiput, and so all round until you meet with the seat of the disorder." Even in recent times the same practice has been recommended by Benjamin Bell, who says we must "form the first perforation in the most inferior part of the cranium in which it can with any propriety be made, and proceed to perforate every accessible part of the skull till the cause of the compression is discovered"

an early period at any rate, an operation of so serious a description, as all recorded experience has shown trephining to be, without more reliable and more clearly-defined evidence of its presence than is commonly thought to denote it. Symptoms which, by the dicta of books, were unquestionably those of compression, have passed off, in the experience of every one, under a treatment of which non-interference was the most important item; while in other cases such large quantities of fluid—blood and pus—have been found, **post-mortem, on the brain, as all** recorded experience tells us *should* have caused a compression which yet never appeared. We find cases on record in which it is evident that traumatic encephalitis was mistaken for compression, and the skull trephined; and in some such instances good effects have followed, evidently from the local bleeding, which, in several of these cases, was considerable; or, perhaps, **from** the preliminary incising of the pericranium, which **we know has,** in some cases, succeeded of itself in removing symptoms analogous to those **caused by compression.***

Blood **rapidly** effused may cause early **compression, which** we know **often** passes off as the **effusion is absorbed; or** mere congestion, the result **of injury, may give rise** to the same symptoms, **and** be allayed **by depletion; yet,** if we **trephine early, we may have only such conditions to contend with.**

---

* **Dease's third case and that of M. A.** Farnham in Guthrie (p. 243 of the last edition) are good and parallel illustrations of this. In both there were signs of pressure on the brain. In Dease's case there were all the signs of pus having formed. In Guthrie's, the paralysis, etc. were the orthodox symptoms of pressure. Both **were** trephined, **and in both** the dura **mater and bone were found perfectly sound. Both were immediately relieved of their symptoms,** and recovered, although **one nearly " died of the doctor." In** the one "the scalping." and **in the other the vessels of the** diploe "bleeding freely," probably account for **the result. These were** both cases of secondary trephining, yet I mention them here with reference to the point hinted at in the text.

If the bone be very deeply depressed on the brain, and
the patient be comatose, with stertorous breathing, slow
pulse, and dilated pupil, then it may be admissible practice
to use the elevator cautiously, with or without the assistance
of Heys's saw; but in all cases in which the bone is not very
deeply depressed, and in which these symptoms are not very
decidedly marked, nor have continued for a considerable
time, I do not believe any interference should be attempted.

It is too much the custom, I think, to deny or overlook
the danger which arises from the operation itself. This is
no place to inquire what is the source of this danger, whether
it be the admission of atmospheric air to the membranes, as
supposed by Larrey and Stromeyer, or the renewed irritation and injury of the brain coverings, or, as others say, from
pus poisoning; but the fact recurs that the most serious,
and at times fatal, symptoms have followed the **operation**
itself in cases in which, **contrary** to expectation, the parts
below the bone were found sound.*

---

* The mortality which attends the operation of trephining needs
little proof, as it is one of the best recognized surgical facts. Take
such a statement as that of Stromeyer, who tells us that during the
three years he attended the hospitals of Vienna, London, and Paris,
he had **not** met with a single successful case, while many severe injuries recovered which were left alone. In the New York hospital
only one-fourth **of their cases** recovered, i.e. eleven cases out of
forty-five. In ten of these the operation was prophylactic, and in
thirty-two therapeutic; three of the former and eight of the latter
recovered. In India I find a record of **four cases of** trephining for
symptoms setting in late, and all ended fatally. In the Glasgow
Hospital register I find no record of a recovery after trephining. In
University College Hospital Mr. Erichsen speaks of four cases of recovery in thirteen operated on, and in the Paris hospitals Nélaton
tells that in fifteen years all their operations of this kind **for trau-**
matic effusion have ended fatally. Mr. Guthrie thinks **the danger**
greater when the operation is performed late. He thinks the sooner
it is undertaken, **if** it is to be had recourse to at all, the better, " believing the violence to be greater when done on parts already in a
state of **inflammation, than** when they are sound." Larrey expresses

Injury of the skull, followed at a late date by compression, is perhaps the most hopeless of all the circumstances in which the trephine can be used, yet it seems that in which it is most properly and incontestably employed. Rigors followed by vomiting, a rapid pulse, stupor, delirium, and palsy, usher in a condition of things which, except in rare cases, is fatal. The longer the time which intervenes before the appearance of such symptoms, the more deadly does their indication appear to be.* It is well known that, in the majority of these cases, the pus is so situated that it cannot be evacuated by the trephine. It is either diffused over the brain between its membranes, or collected in depots deep within its substance, or at parts distant from the seat of injury. In a considerable number of cases, however, it lies superficially, when its formation has been occasioned by a concentrated blow like that of a ball, and may be found collected beneath the place of injury. It is only in these latter instances that any good can be got from the use of

---

himself in almost the same words: "We say, then, that the trepan should be applied when it is decidedly indicated, before the invasion of inflammatory symptoms, which show themselves more or less promptly, according to the idiosyncrasy of the patient, his age, and the cause of the wound; and when it is developed, the operation should be delayed till these symptoms cease. If this second period does not present itself, it is better to abandon the patient, devoted to certain death, than to try a useless remedy which can only hasten his last moments."

* The late period at which dangerous symptoms may be set up, the total absence of any irritation caused by foreign bodies impacted in the brain, which is occasionally observed, are well shown in a case related by M. Mansury in his report on Roux's service during the year 1841. A student, with suicidal intent, shot himself by the mouth. The ball tore the jaw, but there were no head symptoms. On the sixteenth day he was so well as to ask for his discharge from hospital, while on the eighteenth head symptoms set in, and rapid death ensued. The wad and the ball were found in the brain, and yet for a fortnight not the least sign appeared of irritation, or of the presence of such formidable bodies.

the trephine; but such cases are sufficiently numerous in their occurrence to indicate its employment in all instances in which distinct signs of purulent collection set in at a late date. " It is plainly an abscess of the brain," says John Bell; "and as it is an abscess which cannot burst or relieve itself, though the trepan may fail to relieve the patient, yet without that help he will infallibly die." In this is expressed the true reason for its use in these most hopeless cases. It is, in fact, a last resource, which we are not justified in refusing to avail ourselves of.

Besides this, it is also true that, in a considerable number of cases in which the pus has not been found immediately beneath the seat of injury, it has been discovered post-mortem, but slightly removed from it, within the brain substance —so near that very little would have effected its evacuation; and it is also well known that success has followed the bold expedient, first practiced by Dupuytren, of plunging a knife into the brain when the abscess was not found on its surface. The case will end fatally to a certainty, if the matter is not evacuated; and in the event of the attempt failing, such a step, if conducted with proper circumspection, will not add to the gravity of the case. The following case is mentioned, not only because of the late appearance of urgent symptoms, but also because of the position of the abscess found after death, which was situated as above referred to : A private in the 29th was hit by a ball above the eye. The frontal bone was smashed, and the ball was lost apparently in the brain. No head symptoms whatever followed. Some loose pieces of bone were removed, but two parts which were depressed were not interfered with. The antiphlogistic treatment was decidedly maintained. For three weeks no symptoms appeared to create alarm ; at the end of that period, however, a good deal of local inflammation was set up, and the depressed portions of bone, being found loose, were removed. Very little disturbance followed this step, and he was finally discharged, about four months after the receipt of the injury, apparently quite well. A month after dismissal he returned into hospital, complaining of feverishness, headache, and a hurried and excited manner. There was nothing particular found at the seat of injury. The cicatrix was in the same condition as when he left the hospital. The brain-pulse was evident, as it had been since the bone was withdrawn. Coma occurred shortly after his admission, ending in death sixty hours from the first bad symptom. When the head was opened, the hiatus in the bone remained unchanged, only that the edges of the aperture were smoothed and beveled off, and somewhat darker in color than the rest of the calvarium. The dura mater was thickened, but entire, and adherent at the place of wound. The other brain-coverings were highly inflamed, and sero-purulent effusion existed between them. A small abscess was found in the substance of the brain, immediately below the place of injury; and behind this, but separated from it by a thin partition of cerebral substance, was a larger abscess in the anterior lobe of the brain, which communicated with the lateral ventricle of the left side. The small abscess had a distinct sac, but the larger one had not. Dr. Taylor, who reports the case, adds: "These collections of pus might have been of some standing, yet the patient had not a bad symptom up to sixty hours before death." It is very possible that dissipation after dismissal occasioned the sad and fatal result.

A soldier of the Royal Artillery was admitted into the general hospital, on the 15th of November, on account of a shell wound dividing the scalp over the inner and anterior angle of the left parietal bone. He walked to the hospital, assisting a comrade who was more severely hurt than himself; and he complained so little that it was with difficulty he could be persuaded to go to bed. A piece of bone about the size of a shilling was found, on examining his head, depressed to the extent of about an eighth of an inch at the seat of injury. He was purged, put on low diet, and his wound dressed simply. In five days he was allowed to rise

16

and assist in the business of the ward—being put inadvertently, by the surgeon under whose care he was, on full diet
and a gill of rum. No bad symptoms showed themselves
for ten days. His bowels were permitted to get costive.
His wound was nearly closed. On the morning of the fifteenth day from admission he complained of giddiness ; his
pulse was rapid, and his face flushed. Leeches and cold
were ordered to the head, and a purgative administered.
He rapidly grew worse. The wound, now dry and unhealthy,
gave out but a slight gleety discharge. He made many attempts to vomit, which was encouraged by an emetic. His
pupils became widely dilated, but remained sensible to the
action of light. A fortnight after the setting in of these
symptoms he was found to be hemiplegic on the *left* side.
I saw him at this period for the first time. His respiration
was sighing, and numbered twenty-two in the minute. His
pulse was ninety, and contracted. His mouth and tongue
were drawn to the *right* side. He was sensible when roused,
but lay in a half state of sopor when not addressed. The
next day the trephine was applied to the seat of injury, and
the depressed bone removed or elevated. The dura mater
was covered by a pulpy mass of lymph. No pus was found.
Some spiculæ of the inner table which lay on the dura mater
were withdrawn. His symptoms in no way improved. His
tongue was next day drawn to the left side, but his mouth
was unaffected. He had several severe convulsions over
both sides of his body, and he died two days after being
trephined. The skull was found fractured across the sagittal suture into both parietal bones. The dura mater was
little detached round the seat of injury ; but it was there
dark and pulpy, having a semi-organized clot on its surface.
The brain was softened at the place of injury, and had a clot
as large as a walnut lying on it ; while at two points on the
opposite hemisphere, at the edge of the longitudinal fissure,
soft spots were found about as large as a sixpence. Pus
existed abundantly below the membranes, and bathed the

surface of the right hemisphere, as well as extended to the base of the brain, between the hemispheres and under the cerebellum.

The neglect as to diet and the maintenance of the secretions were probably the cause of death in the above case. It is certainly not always easy to maintain as careful a supervision on these points as is necessary, when no functional disturbance whatever is present, and the injury seemingly slight; but this is only one of many examples which might **be** adduced to show the necessity of the long and careful watching which such cases require.

The above was one of the only two instances in which the trephine was **employed** in the general hospital, **and** both ended fatally. In the other case, it was used by one of my colleagues for signs of compression setting in early, with bone much and extensively depressed.

Finally, judging of this question from an examination of the writings of our great masters, the conclusion which presents itself is, that as the symptoms calling for the use of **the** trephine have been so variously interpreted by men of **ex**-perience ; that as the operation has failed as often as it has succeeded in removing the dangers apprehended ; that as the good which has occasionally followed is ascribable, in many cases, to other concurrent circumstances, and not to the removal **of the bone** ; and finally, that as the operation, *per se*, is not devoid of danger,—we should never have recourse to the trephine, unless the indications for its use are very decided, have been present for some considerable time, and have not been assuaged by other remedial measures.

Further, I am disposed, not only from reading, but **also** from the observation of not a few cases which fell **under my** notice during the late war, to conclude, regarding the cases and symptoms which demand operation—that, *primarily*, operative interference (under which term is included the use of the trephine, saw, or elevator) in gunshot wounds of the head should never be had recourse to, except (1) in cases of fracture with great depression—cases in which the bone is forced deeply into the brain, especially if it is turned so that a point or an edge is driven into the cerebral mass; or (2) unless we clearly make out the impaction of spiculæ, balls, or other foreign bodies in the brain, which cannot be **removed** through the wound by means of the forceps: that, *secondarily*, the cases which call for operation are (1) those in which a foreign body is at this period discovered irritating the brain, and which cannot be extracted without a piece of the bone being removed; or (2) those in which signs of compression, set in after a well-marked rigor, continue to increase in intensity, notwithstanding treatment, and have lasted for some time.

In the **treatment of gunshot injuries of the head**, operative proceedings form the least important items, as they can **commonly** be avoided if the rest of the management be judicious, and their success will chiefly depend on a careful attention to less imposing but more important measures.

In their examination the finger should alone be employed, and that even with much caution. They should not be enlarged, unless a more important object be held in view than to clear up doubtful points of diagnosis. If the bone be so extensively destroyed and depressed as to demand early interference, it will make itself sufficiently evident without its being necessary to incise the scalp for the purpose of making the distinction. Stromeyer fitly recommends the application of a piece of wet linen to the wound, which, as it adheres to the scalp, excludes the air. Cold—ice, if possible, or if it cannot be had, simple water—should be applied over this; the patient put to bed in a tent by himself; an active purgative administered, and a most meager diet allowed. The utmost quiet should be enforced, and, in short, the antiphlogistic treatment very decidedly and completely carried out. He should be visited frequently; and if any signs of inflammatory or excited action supervene, instant and copious bleeding should be put in force. "Of all the remedies in

the power of art," says Pott, "for inflammations of mem-
branous parts, there is none equal to phlebotomy, and if
anything can particularly contribute to the prevention of the
ills likely to follow severe contusions of the head, it is this
kind of evacuation; but then it must be made use of in such
a manner as to become truly a preventive, that is, *it must
be made use of immediately and freely.*" I never saw any
good arise from the use of tartar emetic in these cases.
Cold locally, purgatives, low diet, and early bleeding, repeated freely when signs of disturbance showed themselves;
these, with the application of leeches in some cases to the
head, seemed always sufficient, as they are the most useful
means of treating such patients.

As to the extraction of balls when lodged in the brain,
the rule, I believe almost universally followed in the army,
is to extract them if they can be at all got at. It is true
that masses of a far more formidable nature than balls have
remained on and even in the brain without mischief, and
that balls have been discovered encysted years after their
entrance. But these cases form a mere fraction of the number in which the presence of the ball has determined fatal
complications; yet they are the *ignes fatui* by which some
would mislead us from the plain path of duty, which inculcates the removal of such foreign bodies, if at all practicable.
Sir B. Brodie, arguing from an analysis of the published
cases, advocates their abandonment unless superficially
placed; but from this view nearly all military surgeons dis-
sent. In our proceedings, however, "boldness must not
partake of temerity." Few would have the courage or confidence of Larrey or Sir Charles Bell, to follow and extract
the ball from the side of the head opposite to the place of
entrance, or, like Sédillot, pursue it to the depth of several
inches in the cerebral substance; yet all reasonable attempts
ought to be made for its extraction. "Nothing," says Sir
George Ballingall, " will induce me to countenance the practice of leaving it there, except the impossibility of finding

16*

it;" and again, "I am of opinion that it ought to be extracted even at the risk of some additional injury; in short, the prohibition of violence ought rather to apply to the search after balls than to the operation of extracting them." "We have already cited several cases," says Quesney, "which teach us that foreign bodies may remain a long time in the brain without causing death; but with this knowledge we must also bear in mind that it is our duty to extract these bodies, which, sooner or later, almost always prove fatal to the patients; and when we have reason to suspect, from the events, from the instrument which inflicted the wound, or from the state of the fracture of the skull, that such bodies are **retained** and concealed in the substance of the brain, we should make the necessary examinations for the discovery."

If the ball has penetrated deeply into the brain, it is a matter of little moment what steps are taken. Perhaps the best line of conduct is to let the man die in peace. I have never known a case of perforating gunshot **wound of the** head recover. Some such are, however, on record.

Cases in which **pieces of loose bone** remain on the dura mater do not always require to be interfered with. Many surgeons of large experience in the Crimea preferred leaving them to be thrown out by the natural effort, and were not particular even about keeping the wound open. However, I believe this practice to be often dangerous, and that loose portions of bone should always be cautiously removed. The **evil** effects of leaving them, as well as the injurious influence of too early a recurrence to a stimulant diet, were well marked in **the** following case: M'Louchlin, a private in the Connaught Rangers, aged 19, was admitted into the general hospital on the 8th September. He had been knocked down and rendered insensible by a blow from a piece of shell in the final assault on the Redan. A scalp wound, two and a half inches long, was found extending from before backward over the vertex of the head, and a small piece of bone was observed to be depressed at its anterior extremity.

The patient did not become conscious for twenty-four hours after admission. Purging and low diet comprised his treatment. Cold dressing, and nothing else, was applied to the wound. He remained perfectly well, complaining only of slight headache and giddiness, for three months, small pieces of bone being discharged in the mean time from the wound, which had almost closed. After being about a month in hospital, he was allowed full diet, and a gill of grog daily. On the 8th of December, three months after receiving his wound, he complained of a sort of transient paralysis of the left arm, which, although it continued only for a second or two at a time, recurred frequently. His sense of smell, too, suddenly left him. There was no other symptom. On being questioned, he said he had had a rigor and several "fainting fits" during the days immediately preceding that on which he first complained of the paralysis. Next day he had a more prolonged fit of paralysis during the night than he ever had had before, the attack being preceded by pain in the left side. I first saw him during an attack on the 9th of December, which was more severe and more prolonged than any preceding one. His left arm hung powerless, and there was complete anesthesia of the left arm and side from the clavicle to the false ribs, and from the line of the nipple to the spine. The left side of the neck behind the sterno-mastoid was also without sensation. His face was unaffected. The integuments around the wound were puffy, and very sensitive. He said that his uneasy feelings had gradually increased as the wound closed. His bowels were opened freely, and a light poultice was applied to the wound, which was incised. The fit he had on the 9th passed off, leaving the arm weak. The sensibility of the left side slowly returned during the succeeding days. The fits of paralysis came and went, his arm recovering its power, in a great measure, between them. A sharp bit of bone was at last observed lying on the dura mater, and when it was removed, the untoward symptoms disap-

peared. Shortly after this he came under my care. By
quiet and the use of unstimulating food and laxatives he
progressed most favorably; but on several occasions transient feelings of weakness—for there never again was a state
of paralysis established—passed over the left side, when any
scale of bone became loose and lay on the dura mater, and
so soon as this was removed these feelings left.   If his
bowels became costive, even for a very short time, not only
did the headache and giddiness increase, but the numbness
in the side returned.   When he left for England no bits of
bone could be discovered, and the wound was nearly closed;
and he is now, I understand, doing duty with his regiment.
Many of the symptoms in this case were those set down as
calling for the use of the trephine; but the cautious removal
of the fragments when loose, the local bleeding, and the
purging. did all that was required.

Stromeyer warns us particularly against attempting too
soon to remove pieces of necrosed bone, as he thinks they
do little harm if allowed to remain.   In this my own observation leads me by no means to agree.   If the dead piece
can be removed without violence, I believe it should always
be done as soon as it is found to be loose.

On the treatment of hernia cerebri I have no remarks to
offer.

Hardly less important than the immediate treatment of
gunshot wounds of the head, is their after-management.
No class of cases requires more lengthened and careful
supervision.   Relapses may occur long after the patient is
apparently beyond danger; and from the most insignificant
causes—of which, perhaps, irregularities in food, the use of
alcoholic stimuli, and retained evacuations, occupy the fore-
ground—a chronic inflammatory condition of the membranes
is apt to become established, which is no less difficult to
manage than dangerous in its ultimate results.   Very
many cases are on record in which men with balls imbedded
in the brain have apparently recovered completely, but

have suddenly fallen down dead when they had got drunk or excited.

The following cases are added, as in some measure illustrating **injuries of various parts of the head.** They are selected from a large number whose features are nearly parallel :—

Hughes, an artilleryman, was admitted into the general hospital under my colleague, Mr. Rooke, on the 15th of November. He had been struck over the upper part of the occipital bone by a piece of shell, when the siege-train on the right attack exploded. He was rendered insensible by the blow. The scalp was considerably lacerated over the right upper part of the occiput, where a stellate fracture was found, part of the bone being depressed for about a quarter of an inch below the surface. He recovered some degree of consciousness a short time after receiving the blow, but was dull and stupid when admitted into hospital, answering questions if urgently put to him. His head was shaved, and cold applied. The next day he was rational; his eyes were bloodshot, but beyond this there was no bad symptom. Purging, and cold locally applied, were used. A few days afterward he had headache and intolerance of light. Dimness of vision and flushing of the face followed, but there was no notable peculiarity in the pulse or pupil. Leeches were now applied to the mastoid processes; beyond this, the use of laxatives and low diet, nothing else was required to dissipate all threatening symptoms, and he left for England, in January, quite recovered.

In the above case we had merely concussion at first, followed by a threatening of traumatic encephalitis. The treatment was simple, and the cure complete.

Clarke, private, 38th Regiment, aged twenty-two, was wounded on the morning of the 18th of June, but was not brought into hospital till the evening of the 19th, as he lay where he could not be got at till the armistice. A piece of shell had struck him on the upper part of the occiput, laying the scalp open to an extent of two inches and a half. The bone, though denuded, was not seen to be fractured. His symptoms were dizziness, pain in the forehead, and great throbbing in the temples. He was quite rational, but dull, and had double vision and strabismus. His pupils were slightly contracted. His chief complaint then, and for some days after, was of his neck and lower jaw, which had received no injury; but the parotid and submaxillary glands were swollen on the wounded side—a symptom which I have observed in several similar cases. His pulse was forty per minute when lying down, and sixty-nine when he sat up. By active purging, and cupping the nape of the neck, and by the use of low diet, his bad symptoms gradually disappeared. For some days after admission his pulse did not change, except that on one occasion it fell to thirty-eight beats per minute; but as he got better, it rose to the healthy standard. On three different occasions, while he was under my charge, his bowels being unrelieved for a day, his bad symptoms returned in a modified degree, and his pulse sank; while whenever his bowels were freely opened, all uneasiness vanished, and his pulse again rose. The alternation was most curious, and very rapidly developed. This case, like many others, illustrated well the marked sympathy which exists between the head and the bowels. The same slowness of the pulse was noticed by Dr. John Thomson, in the case of a similar injury after Waterloo.

A French soldier received a ball about an inch behind the left ear, which escaped above the eye of the same side. His antagonist, who shot him, was close to him at the moment he fired. This man fell down insensible, and was carried to the ambulance; but he recovered his senses before his arrival there. There was a little blood oozing from both openings; he was dull, but sensible, and complained much of a throbbing pain throughout his head. The ball having escaped, nothing was done for him, further than picking away some small loose fragments of bone, and applying wet dressing. He was freely purged, and got no food. In twenty-four hours, the pain in the head having greatly increased, and being accompanied by delirium, with rapid pulse, ferrety eyes, and hot skin, he was largely bled, and cold was applied to his head. His symptoms were relieved, and from that day he never had a bad symptom, all the treatment his case required being merely low diet and free purging.

Another almost identical case occurred in our own hospital at Scutari, where I saw the patient under the charge of Staff-Surgeon Menzies. The ball had in this case entered two inches behind the left ear, passed deeply, and was removed from the temple. Some hemorrhage set in from both wounds, as well as from the ear, a few days after injury, but it was arrested by pressure. He was dull, and complained of headache for a few days after the occurrence of the bleeding; but by low diet and purging he made an excellent recovery, only that his hearing was destroyed on the wounded side.

A soldier, aged nineteen, belonging to the Second Division, was struck at Inkerman by a rifle-ball, over the vertex of the head to the right of the center line. The ball, passing from before backward, "furrowed" the bone, breaking both tables. This patient declared that he never lost his senses, but felt so weak that he had to sit down. He walked to the hospital, where he was twice bled and actively purged. The bone along the line of the ball's passage, being broken into small fragments, was removed with the forceps, and cold was applied. The brain was bared, but the dura mater, although scratched, was not found torn. A threatened attack of inflammation of the brain was successfully combated by repeated venesections and purging, and the patient made a good recovery—a sulcus about two inches long being felt by the finger over the vertex, the brain pulsations being distinguishable at one extremity of it.

An artilleryman was wounded on the eighteenth of June

by a piece of shell over the back part of the head and rendered insensible. He soon recovered, rose, and walked un-
assisted to the general hospital. No fracture was at first
detected, and the lacerated scalp wound which existed was
dressed simply by the surgeon under whose charge he fell.
Headache alone was complained of for some days, during
which period he was kept low, and freely purged. When
the wound was nearly healed, he was unfortunately allowed
butcher meat and a gill of rum. About a week afterward,
severe cerebral symptoms rapidly and suddenly showed themselves, and the wound took on an unhealthy action. The injury
was now more carefully examined, the scalp being incised to
assist the investigation. A fracture of the occipital bone
was found. Bleeding was encouraged from the incision;
leeches were placed on the mastoid processes; he was well
purged, and cold applied to the head. His diet was again
reduced. The unfavorable symptoms almost immediately
subsided, and, by the use of low diet and purgatives, soon
totally disappeared never to return. In this case a too generous diet doubtless caused the appearance of the unpleasant
symptoms which supervened, and which, if not promptly
arrested, would have been fatal. The local bleeding assisted
materially; but the active purging, the cold applications,
and the low diet were the chief means of saving him.

The following case, the particulars of which were kindly
furnished me by Acting Assistant-Surgeon Brock of the
47th Regiment, was a most interesting one, not only from
the extent of the injury, but from "the phases of recovery:"

Keefe, a private in the 47th Regiment, aged 23, was
struck, on the 15th November, by a piece of shell over the
vertex of the head, and felled to the ground. When found,
a short time afterward, he was apparently dead. The surface of his body was cold, his pupils widely dilated and insensible to light, no respiration or motion of the blood perceptible. His face was much scratched and congested.
Some blood flowed from the right nostril, and the superfi-

cial veins of his neck were gorged. The main wound in the scalp extended nearly from ear to ear, across the vertex of the head; and lesser wounds passed in different directions from this great one. The flaps of the scalp formed by these wounds were reflected in different directions. A large portion of the bone was seen to be destroyed, and the space left was filled by coagulated blood. The patient was seen by several surgeons, and so impressed were they that life was extinct that he was carried to the tent set apart for the dead. Twenty-five minutes afterward, on being again visited, some faint signs of life were observed. There was a flutter at the wrist, and an occasional sigh. Profuse bleeding from the head followed, and on the clot, which was seen to be mixed with cerebral matter, being removed, it was found that the bones forming the vertex of the head were destroyed to the extent of $2\frac{1}{4}$ to $2\frac{1}{2}$ square inches. In this was included part of the superior angle of the occipital bone and a part of both parietal bones, the sagittal suture being clearly defined along the center of one detached piece. **Part** of this extent of bone was altogether gone, and the rest, being detached, was removed.

The surface of the dura mater was scratched, but not torn, except at one spot—at the lateral and posterior part of the wound—where it was lacerated, and from which a spicula of bone an inch long, and which was imbedded in the right hemisphere of the brain, was removed, a piece of cerebral matter the size of a nut adhering to it. The brain at this part seemed soft and broken down. Some depressed bone was elevated, and all loose scales removed. The scalp was brought together by suture, and lint wetted in cold water applied. Next day the patient was quite unconscious, lay on his back, and breathed regularly and naturally. His pulse was very weak, and his surface warm and moist. He passed his urine in bed. His pupils were dilated and insensible to light. He could swallow freely. During the two following days his state was unaltered. Both eyes became

17

affected with strabismus. The treatment consisted of purging, cold to the head, and the most sparing diet. On the
fifth day there were some signs of returning consciousness.
He tried to change his posture, and crossed his arms on his
breast. His pupils, too, acted feebly, and a profuse perspiration covered the surface of his body. On the following day he again relapsed, and the wound, which had begun
to suppurate, now became glazed and dry. When his bowels
were got to act freely, he again improved and became conscious. He complained of pain in the head and down the
left side of his body. Thus he went on till the eleventh
day, being conscious and able to speak. His bowels were
carefully kept acting. His pupils had up to this time come
to contract and expand freely, and the wound was suppurating kindly. He slept much, and expressed a great desire
for food. On the eleventh day he became suddenly restless
and delirious, particularly at night. The strabismus returned. His eye became dull and semi-glazed, and his pupils
were widely dilated and little affected by light. By the
eighteenth day these untoward symptoms had in a great
measure abated. He was sensible, and craved for food.
His left side was found to be paralyzed, the face not, however, being implicated. His pupils were still somewhat
dilated, but active. There was also some œdema of the feet
and ankles. By the twenty-third day, granulations had
formed round the wound. Part of the scalp had adhered
by the first intention. His sleep was now natural and un-
disturbed. Except the temporary irritation caused by some
spiculæ of bone, he went on improving from that time. Attention to his diet and the state of his bowels, and allowing
a free exit for the secreted pus, comprised all the treatment
followed in this case. If his bowels were for a day unrelieved, the bad symptoms immediately reappeared. I examined him previous to his going to England, in January,
and at that period he was in every way recovered. The
head wound was entirely closed, but a depression to the ex-

tent of about three-fourths of an inch existed over the site
of the injury, and the pulsations of the brain were quite
perceptible.

I learn from Deputy-Inspector Taylor that Keefe was
invalided at Fort Pitt, on the 28th May, 1856, on account
of "general loss of sensibility and motion, partial in the
upper, but most complete in the lower extremities." He
was in hospital at Chatham, from 23d March to 26th June,
1856, his state being as follows: "The wound on the head
formed two sides of a triangle, and is about two and a half
inches in length on the right side, and much longer on **the**
left. It is quite healed, but there is a very considerable
depression. The pulsations of the brain are quite perceptible. Complains of severe pain across the forehead, of an
intermittent type. Has lost the power of his lower extremities, with the exception of being able to draw them up and
stretch them out in bed. Has not lost much flesh, and his
general health and functions good." He thus appears **to**
have relapsed after leaving the Crimea, as the marked
paralysis he had at Chatham did not exist when he left
camp.

The intermittent headaches, spoken of in this case, are
among the most troublesome sequences of injuries of the
head. A careful regulation of the bowels and diet, with
blisters to the nape, and morphia, appeared to me the best
remedies. It is a remarkable feature in the progress of head
cases, how often the setting up of subacute inflammation
shows itself by an aggravation of the leading symptom—
whatever that may be — which had existed before: the
headache, palsy, or epileptiform fits. This was clearly
defined in several cases.

The following is an example of **a** severe injury **of the**
fore part of the head, caused by a piece **of stone**:—

A French chasseur-a-pied was struck on the center of
the forehead, above the root of the nose, by a piece of stone
about the size of a walnut, knocked up by a shell. The

stone completely buried itself, and required some skill to extract it. Pieces of bone, comprising nearly the whole ethmoid, were discharged, and a large hole in the frontal **bone** resulted. Three days afterward transient but easily-allayed head symptoms appeared, and he made a most excellent recovery, with a fistulous opening, however, **remaining**. The interest attaching to this case arose from the fact that the inner table of the skull was not fractured, and from the almost total absence of any head symptoms.

It is well known that balls may perforate the outer table of the skull on the forehead, without injuring the inner. Of this the above may be taken as an example, although a stone, and not a leaden ball, was the missile. Several cases occurred in the Crimea of another wound on the forehead **which** is curious, viz., such as are caused by balls passing from side to side of the head below the level of the brain, but destroying one or both eyes.

At Inkerman a French soldier was struck by a ball over the upper part of the left parietal bone. A comminuted fracture was caused, the bone to the extent of a square inch being so broken and detached as to be removed at the first dressing. The dura mater was slightly injured, and a small spiculum, which had been driven into the brain, was withdrawn. He remained speechless for about a week, then articulated hesitatingly, and finally, about six weeks from the receipt of the wound, completely recovered his power of speech. The curious thing in this case was, that perfect anesthesia of the thumb and first two fingers of the right hand existed from the moment of injury, without any loss of motion whatever, and that this slowly disappeared as the wound healed, and he recovered.

To multiply cases would be of little use. The teaching **of all** was to lead us to wait; to purge the patient thoroughly; to remove only such pieces of bone as could be got at with forceps, and which were quite detached and loose; to bleed, if need be, locally and even generally; to use

cold applications when there was a fear of inflammation; to
enjoin perfect rest not only to the body generally, but, if
possible, to give repose to the special senses also, by isola-
ting the patient, and thus removing the stimuli to their
exercise; to enforce the lowest diet, and to continue all this
treatment for a long period, even after all danger seemed
past; and, finally, to treat any incidental complications on
general principles.

It is extremely difficult to get soldiers to avoid stimuli,
or to attend to their secretions; and the desire for improved
diet leads them sometimes to deceive one as to their feelings.
The discipline of a field hospital can often be infringed, and
as it is not easy to persuade men of the soldier's disposition,
of a danger of which their sensations give no warning, it is
necessary to watch them with great care.

Hepatic abscess I saw none of, and the nervous irritation
and weaknes, which so often follow injuries of the head, fell
seldom under my notice, from the transference of the
patients to the rear as soon as their wounds were healed.
Jaundice was present in several fatal cases in which the
head received injury.

17*

CHAPTER VIII.

WOUNDS OF THE FACE AND CHEST.

After the 1st of April, 1855, to the end of the war, there occurred 382 cases
of simple flesh contusions, and wounds of the face more or less severe,
and one death is classed under this head. Of wounds penetrating, or
perforating the bony structure of this region without injuring important organs, there were 107 cases, and 10 deaths; and of those accompanied by lesion of important organs, 44 cases appear, and 3 deaths.
Most of the fatal results were owing to other concurrent causes.

Wounds of the face have been interesting chiefly from
the rapidity with which even the most severe and dangerous-
looking of them heal. The extreme vascularity of the tissues
of the face endows them with a vitality which rectifies most
injuries, and the surgeon is often enabled, both on this
account and from their great distensibility, to repair the
loss which has been sustained, even when that has been
very extensive. It would be much easier to say where and
how the face has *not* been pierced by balls, than to enumer-
ate the directions in which it has. The upper and lower
jaws have been fractured, and large portions of them removed, yet, with few exceptions, a good recovery has
followed, when no other concomitant injury assisted to bring
about an unfavorable issue. One or other of the lower
maxillæ, anterior to the masseters, has been carried away,
and in one case which came under my notice, but which
ended fatally, both lower maxillæ were removed by a round
shot.* The upper jaw has been completely destroyed, and

---

* In a very interesting paper read to the Imperial Academy of
Medicine, by M. Hutin, in April, 1857, there is an account given of
an inmate of the Invalides, (to which M. Hutin is surgeon-in-chief,)

in one case which occurred in the 31st Regiment, a grape-shot, seventeen ounces in weight, was impacted in the superior maxilla, and necessitated the removal of most of the bone.

Hemorrhage is undoubtedly the great source of annoy-ance and danger in gunshot wounds of the face. The difficulty of commanding it is at times so great as to place the patient in imminent danger. It frequently appears early, but stops spontaneously. Men who have received a severe face wound seldom leave the field without sustaining a considerable loss of blood, and secondary hemorrhage is common when the bones have been fractured. The depth, irregularity, and extreme vascularity of the parts make the application of a ligature to the bleeding points difficult, and to be effectual, compresses must be applied with much niceness. It is in wounds of the deep branches of the face, in which secondary hemorrhage has taken place from a

who had the lower jaw carried away by a cannon-ball at the battle of Wagram, forty-eight years ago. He recounts the changes which the parts have undergone since. It seems that the hemorrhage was very severe at the moment of injury, but that it ceased sponta-neously. The tongue hung down in front of the neck, and was never drawn into the throat—an accident which did not occur in four other cases, in which M. Hutin has known a like injury produced by a like cause. The patient referred to by M. Hutin has worn a silver mask since his accident, which protects his tongue hanging out, and ad-herent as it is to the neck. By means of this mask the variations of temperature do not affect the wide void which exists in the floor of the mouth. The most remarkable change which the progress of time has brought about in the parts is, that the upper jaw, in place of preserving its horseshoe shape, has become so contracted at its middle as to assume the figure of an hour-glass. This change began to take place three years after he was wounded, and has gone on in-creasing up to within a short time. The secretion and loss of saliva is great, but the patient enjoys perfect health. There is an interest-ing question raised by this case, viz., whether an analogous change may be looked for in those instances—of late years pretty numer-ous—in which the lower jaw has been excised.

sloughing surface, that Anel's operation, performed on the main artery, may be said to supersede, from necessity, Bell's doctrine of local deligation.

The branches of the facial nerve are sometimes so much injured in wounds of the face, either by the ball or by the fractured bone, that temporary and even permanent paralysis may ensue ; but there is one source of danger in these cases which does not always obtain the attention its importance demands. I refer to the swallowing of the secretion from the wound. If great care be not taken to remove all the morbid secretion which results from injury of the bones of the face, if any amount of it gets into the stomach, much constitutional irritation will result, and a fever of a low typhoid and very fatal form will be caused. I believe I have seen this result very clearly follow the cause referred to in some cases. In one case, where a sergeant of the Buffs died in the general hospital from the effects of a severe face injury, by which the anterior part of the lower jaw and a small portion of the upper were fractured by a round shot, I suspect the fatal result was at least accelerated by the cause mentioned, although the utmost care was taken to prevent its occurrence. He was a very unhealthy man, who had just been discharged from his regimental hospital a few days previous to the accident, and was of a nervous, irritable disposition. He was struck from the side by a small round shot, which had previously struck the parapet of the trench. The symphysis and part of the body of the lower maxilla, as well as a small portion of the upper jaw, were destroyed. The soft parts, especially at the chin, were much torn and bruised, and ultimately sloughed. When examining his chest, on account of a cough which troubled him on admission, a cavity was discovered in one of his lungs. Hemorrhage took place repeatedly from branches of arteries opened as the slough separated. By maintaining an opening below the chin, and washing the wound from the mouth, the greater part of the abundant secretion was removed ; but yet no small quantity found its way backward into the throat, and was swallowed. His stomach became very irritable, his strength failed, and a low muttering delirium preceded death. A putrid abscess occupied the summit of one lung, and pus was infiltrated among the tissues covering the trachea.

In fractures of the bones of the face from gunshot, we make an exception to the general rule of removing fragments which are nearly detached. The large supply of blood which is sent to every structure in this region enables pieces of bone to resume their full connection with the other tissues, when detached, in a way that would be fatal to similarly placed portions in other parts. Hence the rule, not to extract any spiculæ whose attachment has not been completely destroyed, and whose direction is not opposed to a proper union of the broken parts. The exfoliation which follows in injuries of the bones of the face is slight as compared with those of other parts.

The destruction or injury of one or other of the organs of special sense, and the deformity which may be caused, as well as the tedious exfoliations which at times follow severe face wounds, are the chief ulterior causes of suffering and annoyance to which they give rise. In cases in which the lower jaw is destroyed, the loss of bony substance, the powerful action of its muscles, which is so difficult to counteract, and the imperfect mode of repair, contribute to occasion a considerable amount of deformity. It is a sufficiently old though not always remembered maxim, to extract by the mouth, whenever practicable, all balls lodged in the face.

The curious manner in which balls may be concealed in the bones of the face, and be discharged of their own accord, was shown in one instance in the Second Division, after the battle of the Alma. A round ball had entered close to but below the inner canthus of the eye, and being lost was not further thought of. The wound healed, and the patient had almost forgotten the circumstance, when, after suffering slightly from a feeling of dryness in one nostril, the ball fell from his nose, to his great alarm and astonishment, several months afterward. It is somewhat singular that **so** little trouble should have been occasioned in this case, as it not uncommonly happens that a most distressing fetid suppuration attends the injury of bone in the region where this ball was probably lodged.

It is in **wounds of the neck** that the extraordinary manner in which the great vessels escape a ball's passage becomes most evident. Thus the neck has been injured by gunshot, more or less severely, 128 times, and yet only 4 deaths have resulted from these wounds. Yet it must be true that a large number die on the field from these injuries. It would be useless, but sufficiently easy, to record cases in which balls, and even bayonets, have traversed the neck, and yet did not injure the great vessels; sometimes passing from side to side, sometimes from before backward, it would appear almost impossible that the blood-vessels could have escaped the wounding agent, and yet no indication of any mischief followed. The great nerves suffer not uncommonly in gunshot wounds of the neck, when such wounds **are** situated low down. Paralysis of the arm setting in, either immediately after the infliction of the injury or a few days later, affords evidence of such a lesion.

The **soft coverings of the chest** were wounded, after April 1st, 1855, by gunshot, more or less severely, 255 times, with 3 deaths resulting. In 24 cases, the bony, cartilaginous, or intercostal tissues were wounded, and one of these died. Lesion **of the contents took** place 16 times, although the ball did not penetrate, and 9 deaths resulted from that cause. The ball penetrated and logded, or appeared to lodge, 33 times, and of these patients 31 perished, while in 9 cases the contents of the thorax were wounded superficially, 3 times with a fatal result. In 83 instances the con-

tents were deeply perforated, and death followed in 71 cases.*
It would thus appear that, with all our boasted improvements
in the method of investigating the effect and progress of injuries of the lungs, the mortality has not abated much from
what it has always been, when large numbers of men have
sustained such injury from gunshot. **Wounds of the thorax**
are very common in battle when the combatants are in close
proximity. This was particularly the case in the civil dis-
turbances in Paris; and in siege operations the same holds
good. The large surface and elevated position of the thorax
accounts in some measure for this.

The distinction usually made between **wounds of the
parietes** and those which penetrate and injure the **viscera of
the cavity** is evidently a good one, as it separates between
two classes of injuries of very different import.

Simple contusion of the walls may be caused by a spent
ball, or by a ball which has impinged against some part of
the soldier's accoutrements, and has thus been prevented
from entering. Such an injury, although not accompanied
by any fracture, may yet be sufficient to give rise to hæmoptysis, severe constitutional shock, and internal inflammation.
If the ball strike the edge of any of the metal plates which

---

* M. Legouest mentions, in a communication he has been good
enough to send me, 6 cases of penetrating wounds of the chest, as
having occurred in his division of the Dolma Batchi hospital at Constantinople, and of these the half died. Alcock gives 1 to $1\frac{7}{29}$
as the mortality attending his cases of penetrating and perforating
gunshot wounds of the thorax. In Guthrie's 106 cases, of whom a
half perished, "the cavities were not penetrated." In the documents
of the medical department I have found a record of 39 cases in **which**
the chest was penetrated, and in some perforated by balls. **In most**
of these there were signs of injury to the contents. Of these 39 cases,
27 died and 12 recovered. Menière reports 20 **cases of** perforating
wounds by gunshot, many of them effected **at** very **short** range. All
died, many very soon after being wounded. Nine penetrating wounds
which he also mentions recovered, in all which there were signs of
lesion of the lungs.

form part of the soldier's accoutrements, then the injury to
the contents may be inflicted by the part so struck, as was
the case in the following instance, in which a round shot
was the missile, and the severity of the injury was little
evidenced by the symptoms before death : Darling, private,
61st Regiment,* was hit at Sadoolapore by a round shot,
on the edge of the breast-plate, which was so turned inward
as to fracture the cartilages of the fifth, sixth, and seventh
ribs on the left side, close to the sternum. The skin was
not wounded. He walked to the rear, and complained but
little for two hours, when he was seized with an acute pain
in the region of the heart. His pulse became much accelerated, and he grew faint and collapsed. A distinct and
sharp bellow's-sound accompanied the heart's action. He
died in seventy-two hours from the receipt of the injury—
the pain and dyspnœa, which had been so urgent at first,
having abated for some hours before death. The heart was
found to have been ruptured to an extent sufficient to **allow**
of the finger being thrust into the left ventricle. The obli-
quity of the opening had prevented the blood escaping into
the pericardium, which contained about two ounces of dark-
colored serum.

Dupuytren has drawn attention to the long period which
ball wounds of the soft parietes of the chest take to heal,
especially when they are " en gouttière." This he accounts
for by the constant motion imparted to the walls by the
movements of respiration.

If the blow from a ball be forcible, or strike directly on
the chest without the intervention of any strong substance,
then fracture of one or more of the ribs will probably be
caused, and possibly pleural or visceral inflammation as
well, from the effects of the blow, or the presence of spiculæ
driven inward. These fragments are at times long and
sharp, and may be totally detached from the rib, and carried

---

* Unpublished records of the Medical Department.

deeply into the lung substance. The cartilage of a rib, although torn by a ball, is seldom driven into the parenchymatous tissue, but remains so attached that its fragments can be easily restored to their proper position.

It occasionally happens that a ball is arrested between two ribs. This happened in the following case: Cassay, a private in the 38th Regiment, was admitted under my charge, into the general hospital, on the 18th of June, suffering from a gunshot wound of the left side of the thorax. The ball, a large conical one with a broad base, was much spent when it struck him. It did not force itself into the cavity, but lay wedged between the cartilages of the second and third ribs, on the left side, about an inch from the sternum. On withdrawing the ball, the cavity of the chest was found to be fairly opened, and the lung was visible as it expanded and contracted. The patient had a severe attack of pleurisy a few days afterward, for which he was repeatedly bled. Effusion, to a limited extent, followed, and his gums were touched with mercury. For five weeks the wound continued to suppurate freely. The lung became adherent to the parietes. This patient had subsequently **a** short attack of bronchitis, but ultimately made a good recovery. He went to England in August, at which time he still complained of a severe pain in the left clavicle and shoulder, which extended down to his hand, and was attended by numbness and want of power. The pain was increased by touching the arm, and had continued since he was wounded. In this case the cavity was opened, but the lung escaped injury. The non-collapse of the lung was well seen in this, as in some other instances which fell under my notice. **The** natural mode of repair, by adhesion between the lung and the walls of the chest,* and the troublesome affection arising

---

* The advantage of this adhesion of the pleuræ, and the part which it plays in the repair of chest wounds, is well brought out by Roux in his Mélanges de Chirurgie.

from injury to the nerves of the arm, were both illustrated in
the above case.

Pieces of shell not unfrequently open the cavity, but
spare the lung, while sometimes the reverse happens, and
the lung may be injured without the pleural sac being
opened. The following was a curious instance of this latter
accident, without the thorax being opened. The case occurred under the charge of my friend Mr J. H. Hulke,
assistant-surgeon to King's College Hospital, to whom I am
indebted for the details: Private Jeremiah O'Brien was admitted into the general hospital on the 15th November, 1855,
having been wounded by a piece of shell when the right
siege-train exploded. His left arm and forearm were extensively shattered, and he had two small irregular wounds
on the left side of his chest, one just below the lower angle
of the shoulder-blade, and the other on the same level, but
about two inches nearer the sternum. His breathing was
quick and labored, and bright florid blood was bubbling
from his mouth. His face was pale, his pulse flickering and
very feeble. He spoke with a firm voice, and begged his
arm to be cut off. No communication could be detected
between the wounds on the chest and the cavity within, but
two ribs were found to be broken. His wounds were dressed
simply, and his chest fixed. Beyond dressing, nothing was
done to the arm, as he was not in a condition to undergo
any operation. By night the breathing was easier, and he
brought up less blood. Next morning his pulse was fuller,
but intermittent. His spit still contained blood. His chest
was naturally resonant as low as the fourth rib, but below
this, by percussion and auscultation, dullness and friction
sounds were discovered. He was cheerful, but, as he had
not slept, half a grain of morphia was administered. He
subsequently rallied somewhat, but died suddenly next afternoon, without any return of the bleeding. On examination
after death, the sixth and seventh ribs were found fractured
without displacement. The pleura costalis was entire. The

part of the lung below the level of the fracture was entirely adherent to the ribs and diaphragm, while in the upper part of the pleural sac a small quantity of bloody serum was found. Opposite the position of the fractured ribs, the lung substance was extensively lacerated. A large rent ran inward from the external surface toward the root, downward toward the base, and upward toward the apex. A large branch of the pulmonary artery was seen with an open torn mouth in the rent, while many other vessels stretched across it. The right or uninjured lung was ecchymosed at numerous spots on its surface, and in part emphysematous. Ecchymosed points were seen also on the surface of the heart and pericardium. The mitral valves and endocardium of the left ventricle were of a rosy hue. The segments of the tricuspid valve were bound together by a fibrinous clot, which narrowed the passage to the size of a small quill. Blood was found in the small intestines, but not in the stomach. Mr. Hulke remarks the arrestment of the bleeding by the mode in which the chief vessel was torn, as well as the conservative act of shutting off the rent in the lung, and the torn bronchi from the pleural sac by the formation of adhesions.

It is seldom that a conical ball will be found to lodge in a rib, as a round one has been seen to do, or yet to run round under the integuments, or at all to lodge within the chest. In fact, it very rarely fails to penetrate deeply, or pass quite through the entire cavity.

Non-penetrating wounds are more dangerous at some points of the thorax than at others. Thus, when a **ball** strikes a large bone like the scapula or the spine, or in those places where the large blood-vessels and nerves are situated, as in the axilla and upper part of the chest, the **danger is** greatly increased.

The gravity of **penetrating wounds** depends very much on their direction and their point of entrance, as when, with an incidence very oblique to the surface, they enter at some

parts of the chest, they may traverse a portion of the cavity
without touching the contents. So it happened in the following case : Fontaine, a private in the 90th, wounded on
the 8th September, was admitted into the general hospital
on the same day. The ball, after passing through the flesh
of his left arm, which was at the moment in advance of his
body, had entered the thorax in the axilla, and escaped at
the inferior angle of the scapula, fracturing it, along with
two of the ribs, at the place of exit. No immediate disturbance followed, but in twenty-four hours signs of acute pleurisy appeared, and required decided treatment. The ball
had entered the cavity of the chest, but the substance of the
lung had evidently escaped. Bone exfoliated by the wound
of exit, which continued to suppurate long after that of
entrance had closed. No bad symptom arose after the
attack of pleurisy above referred to was subdued. I have
seen this man lately in perfect health.

The finger is the only probe permissible in examining
wounds of the thorax. If we thereby discover the projection inward of fragments of a rib, or portions of it impacted
in the lung, we should take immediate steps for their removal,
even though the wound has to be enlarged in order to allow
of its accomplishment. The ribs are best fixed, and the
wound left free, by means of strips of adhesive plaster
passed from the spine to the sternum, and from above downward, so placed as to embrace the wounded side only. Men
wounded in the lungs require all the breathing space we can
give them, and this is best managed by having the sound
side free.

It is a singular circumstance connected with wounds of
the walls of the thorax, that an intercostal artery is seldom
opened. I neither saw nor heard of such a case during the
war, so that we were spared the adoption of any of those
operative procedures for its closure, which, Boyer remarks,
are more numerous than the authentic cases of the occurrence of the accident.

Balls passing in front of the chest from side to side may cause very grave injury to the parietes, without absolutely wounding either the heart or lungs. This occurred in the following most interesting case :—

Fleming, a private in the 18th Regiment, was admitted on the 18th of June into the general hospital, under Mr. Rooke. This lad was struck by a Minié ball, a little above the right nipple, as he stood sideways toward the enemy. The ball escaped below the left breast. The sternum was fractured and comminuted by the ball in its transit. Severe dyspnœa followed, together with a slight attack of hæmoptysis. Repeated attacks of inflammation occurred over parts of both lungs, and the subsequent supervention of pericarditis necessitated bleeding and the use of tartar emetic, and subsequently of mercury, so as to touch the gums. The soft parts between the wounds of entrance and exit sloughed, and the sternum, to the extent of about one and a half inches, together with the cartilaginous ends of the ribs thereto attached, came away in fragments, or were absorbed, so that by the 12th of July a profusely suppurating wound had formed, 6 inches long by $2\frac{1}{2}$ broad, across the front of the chest, laying open the anterior mediastinum, together with the right thoracic cavity, the opening into which was, however, sealed by the adhesion of the lung to the parietes. At the left extremity of the wound, and at its lower part, the heart was plainly felt only covered by the pericardium. A to-and-fro sound accompanied the motions of the heart, but these were not sufficiently pronounced to prevent the recognition of the two natural notes. Hectic fever, harassing cough, and emaciation supervened. By the middle of July the wound had begun to granulate, and the patient seemed to improve. An attack of diarrhœa, however, prostrated his little remaining **strength,** and ultimately proved fatal. Before death, the **pus with** which the wound was filled receded on inspiration, and welled up when the lungs were emptied, as if it sank between the lungs when

18*

they expanded. On the morning of the day on which he
died, a new sound was heard to proceed from the region of
the heart, to which we never before heard any similar. It
was exactly like the "clanking" note which accompanies
the working of a pump when its gear is loose. There was
the sucking in and expulsion sound, together with this sharp,
peculiar note, which it is impossible to describe, but which
immediately suggested the probability that the pericardium
had been opened, and that the pus which filled the wound
was alternately being sucked into and ejected from its cavity.
On examination, this view was confirmed, as a small hole was
found at the inferior and left lateral aspect of the wound,
through which the pus appeared to be drawn in and thrown
out during the action of the heart. After death it was
found that this aperture led into the pericardium, which was
much thickened, and adherent to the heart for a space of
two inches by one, at the anterior and middle part of that
organ. The opening mentioned led into a pouch formed by
the pericardium round the roots of the great vessels, and
which pouch communicated freely on the right side of the
heart with the sac of the pericardium, at the base of the
heart below the adhesion. Pus was freely effused into the
pericardium, and the surface of that membrane, as well as
that of the heart, was of a drab color and thickly coated
with lymph of a low type of organization. The heart itself
was healthy. The lungs were somewhat congested, and
their anterior surfaces were adherent to the parietes. The
coats of the stomach were unhealthy, but beyond this nothing
was observed.

The noble struggle made against death by this poor boy,
the very extensive injury, the opening of the pericardium,
and the sealing of both sides of the thorax by the pleural
adhesions, were all points of much interest and no little
instruction.*

---

* John Bell (second Discourse on Wounds, p. 302) refers to a case
related by Galen, in which part of the sternum was removed, the

The two following cases show how small a difference in the place of transit of the ball may determine the question of life or death: A Zouave was struck at the Alma by a round ball, which entered the parietes close to the right nipple, and escaped at a corresponding point on the left side. The ball passed in front of the sternum, which it fractured. Curiously enough no inflammation whatever of the contents of the thorax followed, and he was in a short time discharged well. The points of entrance and exit differed little in this and in the case of Fleming, but the projection of the sternum being less in this patient the result was very different.

A Russian soldier lay close to the Zouave just referred to, who, in the same battle, had been struck by a ball about a quarter of an inch to the outside of the right nipple. The ball had then passed behind the sternum, fracturing it badly in its course, and escaped close to the left nipple. Double pneumonia and pericarditis followed, and he died. The whole contents of the thorax were found implicated in one vast inflammation. Not being present at the post-mortem examination, I did not learn how far the pleuræ or pericardium were injured (as I understood they were) primarily.

When a ball fairly enters the chest, and either penetrates or traverses the lung, the danger is most imminent. These injuries, however, are not so fatal, on the whole, as similar wounds of the head or the abdomen. The younger Larrey* and Menière both record the circumstance, that the majority

---

pericardium opened, and the man cured. He thus comments upon it: "Here, then, we have, upon that authority which has been always respected, a case exceeding in the miraculous all that has ever been recorded by the patient Vander Wiel, or gathered by Schenkius, or any German commentator among them—a man with a slow suppuration, confined matter, a carious sternum, and the heart absolutely exposed and bare." In Fleming's case we had all the unfavorable symptoms, but, unfortunately, not the recovery.

* Relation Chirurgicale des Événements de Juillet, 1830, a l'hôpital militaire du Gros-caillou.

of the killed in the civil commotions of 1830, in Paris, succumbed from penetrating wounds of the thorax. The immediate danger will depend upon the depth of penetration,
and the part implicated. If the heart or great vessels are
wounded, death will in general be instantaneous.* When
the lung is only superficially wounded, then the vessels which
are injured must be of small caliber; but the deeper the
ball penetrates, the larger are those encountered, and, consequently, the more mortal is the wound. The patient may
be suffocated at once by the blood, or it may escape in such
quantity as to cause death, within a short time, by exhaustion. If the wound be at all severe, the shock is very great,
and blood generally passes from both the mouth and the
wound. That from the mouth is frothy, while that from the
wound is darker colored in general. The wound being high
in the walls of the thorax will make the escape of blood by
the orifice less in quantity than if it be situated low down,
and such a situation will render the evacuation of the effused
blood or serum more difficult afterward. Air as well as
blood will generally escape by the wound, and thus the presence of these two signs—blood by the mouth, and blood and
air by the wound—are unequivocal proofs that the lungs
have been injured, although their absence does not prove
the opposite.†

The dangers which attend a penetrating wound of the
lung are thus, primarily, hemorrhage and collapse, as well
as those from suffocation, if the bleeding be profuse. The
hemorrhage and the fainting are, by a sort of paradox, both
the patient's danger and his safety. Secondarily, the danger

---

* In the *New York Journal of Medicine*, vol. xiv., there is a very
interesting paper, by Dr. Purple, on wounds of the heart. He makes
reference to several cases in which balls have remained long imbedded
in that organ.

† In the accounts given us of the spear wound which so nearly
deprived Alexander the Great of his life, in the battle with the Malli,
we are told that he blew both air and blood from his wound.

of such wounds proceeds from inflammation and its products, the exhaustion which attends prolonged exfoliations and suppuration, together with that which arises from the organic diseases that are thereby so apt to be engendered.

A short, tickling, harassing cough, attended by bloody expectoration; a cold and bedewed surface; a pale, anxious face; a weak, trembling pulse; palpitations of the heart; oppressed breathing, arising in the first instance, according to Hunter, from the pain occasioned by the action of the wounded lung and muscles, and afterward from the inflammation and effusion,—these are the usual symptoms which attend penetrating wounds of the lungs. At a later date, if the bleeding cease — a circumstance which will be evidenced by the disappearance of the collapse, the return of the heat to the surface, and of strength to the pulse, as well as by the length of time which has elapsed since the infliction of the wound—then those symptoms which result from inflammation appear. We have thus two stages or periods which demand separate attention in our treatment — that during which there is internal hemorrhage with collapse, and that which follows and is accompanied by reaction and inflammatory action; to these I might also add that of convalescence.

The collapse which follows penetrating wounds of the lung, though dangerous, is yet, if not very profound or prolonged, the best guarantee for the patient's safety. To such cases the observation of Hewson is peculiarly applicable: "Languor and faintness, being favorable to the congelation of the blood and to the contraction of the bleeding orifices, should not be counteracted by stimulating medicines, but, on the contrary, should be encouraged." With our modern notions on bleeding, it is often difficult **to reconcile** the necessity, which experience shows **there is, for** energetic depletion when reaction sets in. **The m**ajority of our patients were certainly not subjects in which this remedy could be pushed so far as Guthrie and Hennen would appear to recommend; but I think it was very generally observed that those cases did best in which early, active, and repeated bleedings were had recourse to. It is well known that in sieges generally soldiers do not show their usual tolerance of bleeding, and when their health is so much undermined as it was at Sebastopol, the surgeon is often placed in a most unpleasant dilemma. That many most excellent recoveries were made without having recourse to the lancet, is undoubtedly true; but not a few, I fear, died from want of it. When the loss of blood by expectoration and by the wound has been very free, of course the necessity for abstracting it otherwise will be much less. The system is then far more easily reduced to that point which favors the formation of the "caillot tutelaire." We must, in cases where venesection is required, be especially careful to bleed by a large orifice, and be guided by effects.* This, with perfect rest, the lowest diet, cooling drinks, and possibly digitalis, must form our means of managing the early stage. Any return of the oppression will show the necessity for further depletion. In wounds from gunshot, the patient should be allowed to lie in the position which he chooses; but if the wound be a stab, the position prescribed should be that which will favor the adhesion of the pleuræ; and when there is effusion within the thorax, that which will allow of its escape.

To determine whether the blood which flows from a wound in the thorax proceeds from a wounded intercostal or from

---

* "Until the danger of immediate death from hemorrhage is over," says Hennen, "we must not think of employing anything except depletion by the lancet; it, and it only, can save the life of the wounded man." "It is only by these repeated bleedings," says John Bell, "that the patient can be saved. The vascular system must be kept low in action, and so drained as to prevent the lungs from being oppressed with blood. One thing is very clear," he adds, "that if the surgeon bleed only when the cough and bleeding from the lungs return, he never can do wrong."

the **lung**, has called forth more acumen and research than it would appear to merit. The difficulty will be greatest when a knife has been the instrument, and the wound made is very oblique. In large wounds, Sanson lays down the following means of diagnosis: 1. Whether the blood be arterial or venous. 2. By turning out with forceps the lips of the wound, and seeing whether the blood proceeds from one of these lips. 3. By compressing the superior lip of the wound with the finger, *i.e.* pressing upon the inferior border of the upper rib, where the wounded intercostal may be placed. He objects to the use of a roll of card introduced in the shape of a gutter, because when that can be done we may be able to see the wounded vessel with the eye; but the examination **of** the wounding instrument will often show whether it could penetrate deep enough to injure the lung.

Bleeding from the lung makes itself apparent by both rational and physical signs. Some of these are common to all hemorrhages, external or internal, while others are present in intra-thoracic effusions of whatever description. Of the **rational signs,** paleness of the face, coldness of the surface, a small, concentrated, and quick pulse, giddiness, and syncope are those referable to the loss of blood; while the dyspnœa,* sometimes amounting almost to suffocation, the feeling of weight in the chest, the anxiety, restlessness, and the decubitus on the wounded side belong to all effusions. The **physical signs** are also common to all effusions. They are—a dilated chest, little moved during respiration, bulged intercostal spaces, dullness on percussion, and the absence of vesicular breathing. If there be air also present, we will have added those signs which are peculiar to such a complication, and which are recognizable by percussion and auscultation. The peculiar ecchymosis described by Valentin, and which results from the escape **of** blood into the sub-

---

* Sabatier mentions having seen patients perish of hemorrhagic effusion in whom the breathing was not disturbed, and who could lie **in** any position.

cutaneous cellular tissue, seldom appears; but if it does, it
is according to many a valuable sign of hemorrhagic effusion.*
If, then, after a gunshot wound of the thorax, we have those
signs present which would indicate the loss of blood, as well
as those which indicate the existence of fluid in the pleura,
embarrassing the functions of the contained viscera, the
diagnosis is plain. If blood escape by the external wound
during respiration, or after a cough, the opinion will be
strengthened that blood has been poured out, and occupies
the pleural sac.

The danger from hemorrhage is greatest during the first
twelve hours, and is pretty well over by the second day. A
flow may, however, continue, in greater or less quantity, for
eight or ten days, but then it is seldom to any serious
amount. If the quantity of blood effused be small, it will
probably be absorbed; but if it is in large quantity, and
especially if air is also present, the gravity of the lesion is
much augmented. So soon as all fear of a renewal of the
bleeding is over, the effused blood, if in quantity, should be
evacuated by operation; but, as Sanson says, it is better to
be a little late than too early in taking this step.†

---

* Luez remarks upon this point: "Valentin pretends that the
ecchymosis which is observed on the loins, in wounds of the thorax,
is a pathognomonic symptom of effusion into the pleura, and that its
absence is a counter-indication to paracentesis. Larrey says he constantly observed this fact, as do many other practitioners, such as
Louis, David, etc. However, after the observations collected by De-
granges, Chaussier, Callisen, Saucerotte, and others, we cannot look
upon this phenomenon as a certain sign of hemo-thorax; because, in
many circumstances where the effusion really exists, it has not been
observed, and it has followed non-penetrating wounds."

† "Au reste cette indication n'est que d'une **importance** tout-a-fait
secondaire quand on la compare a celle qui **prescrit** d'arrêter a tout
prix l'hæmorrhagie; aussi avant de pratiquer une nouvelle ouverture
ou d'agrandir celle qui existat deja, convient-il de s'assurer si
l'ecoulement du sang hors du vaisseau divisé a cessé completement.
Hors de cette condition, l'operation n'aurait d'autre resultat que de

There is no question connected with wounds of the chest so difficult to solve as that which has reference to the management of **internal hemorrhage**. The embarrassed state of the lung demands the evacuation of the fluid; and yet, if we allow it to escape, the bleeding from the lung is renewed, and death results. So it was in the following case:—

Hannihan, a private in the Royal Irish Regiment, was admitted into my wards in the general hospital on the 18th of June. While lying on the ground, with his head toward the enemy, he was struck above the left clavicle by a rifle-ball, which traversed his lung from its summit to its base, and was found lying quite superficially in the left lumbar region, from which position it was removed. The dyspnœa, on admission, was very great, and the hæmoptysis most profuse. The surface was cold and bedewed with cold perspiration. The pulse was weak and tremulous, and the decubitis was on the wounded side. The removal of the ball was followed by a tremendous gush of blood from the incision made, and the blood continued to flow in such quantity that I had to close the wound to prevent immediate dissolution. The necessity of guarding against a suddenly fatal event was for the moment paramount to the indication of freeing the embarrassed lung of the effused blood; and as the hemorrhage, moreover, appeared to be active, I wished to try to check it by the pressure which would result from the blood being allowed to accumulate in the thoracic cavity. The patient was twice largely bled, and he had acetate of lead and opium given him. These measures appeared to afford him some relief. Next day he had rallied considerably. His pulse was better, and his look was less distressed. By the afternoon of that day the dyspnœa became so urgent

---

favoriser la continuation de l'hæmorrhagie, **en privant** la plaie du vaisseau de la compression salutaire qu'exercent sur elle le sang retenu dans la poitrine ainsi que les caillots qui ont pu se former."— SANSON, *Des Hæmorrhagies Traumatiques*, p. 260.

that I allowed a considerable quantity of the collected blood
to escape. This gave him for a time decided relief. The
severe exhaustion which, however, soon followed this step,
and the return of the dullness on percussion to its former
level, seemed to intimate a renewal of the hemorrhage;
hence I did not reopen the wound, but determined to abstain from all interference till the bleeding vessel had had
time to close. The patient was so completely prostrated
by the hemorrhage which had evidently taken place internally that I could not have recourse to any further depletive
measures. The stethoscopic examination of the chest discovered amphoric breathing over the upper part of the left
lung, while over the whole surface of the right chest the respiration was harsh and loud. Dullness existed on the left
side from the base of the lung up to an inch and a quarter
above the level of the nipple. There was suppression of
urine for thirty hours after admission. This patient died on
the fifth day, without any change in his symptoms from
those noted above. The left side of the thorax was found
more than half full of blood, for the most part fluid. The
lung was half solidified, and compressed against the spine.
Lymph was effused to a limited extent on its surface. The
ball had traversed the lung in a direction from above downward and backward. Its track was ragged and coated with
lymph. The three upper and the three lower ribs were
fractured. The patient's back, on the wounded side, was
ecchymosed before death, and gave him much pain. This
discoloration bore much resemblance to that ecchymosis
described by Valentin—only it appeared at too early a period, and was not sufficiently pronounced to accord with his
description.

I am not in a position to determine whether the retention
of the blood in the cavity can really exert so great a pressure on the wound in the lung as to arrest the bleeding;
but such was the opinion of Valentin, Larrey, Sanson, and
Dupuytren. I am disposed to think that, in such cases as

the foregoing, it would be better practice to open the cavity freely by enlarging the wound, so as to allow the blood to escape freely, and thus favor the contraction of the lung and the closure of the vessel ; but in Hannihan's case such a step would have been attended with much danger, from his great prostration.

If the lancet be employed in such cases, it is a matter of the greatest nicety, and requires the utmost discrimination and judgment to abstract exactly the quantity of blood requisite for producing the desired effect without exhausting the patient, whose system has been already so much drained by the internal hemorrhage.

Hæmoptysis does not always occur in penetrating wounds of the lungs, and dyspnœa may be but slightly marked at first. The following case was an example of this : M'Kennah, private 77th Regiment, was admitted into the general hospital July 27th. When in one of the advanced trenches, a Minié ball struck him obliquely from the left side at **the** middle of the supra-spinous fossa of the left scapula, and lodged. On admission, a couple of hours after the receipt of the wound, slight dyspnœa was the only observable symptom, and the only thing the patient himself complained of. The finger passed into the wound showed the direction of the ball to have been toward the center of the body, but nothing was detected except some roughness along the posterior border of the scapula. In the evening the dyspnœa was more marked, and the pulse had increased in frequency. The decubitis was dorsal throughout. Emphysema appeared over the surface of the right side of the chest. He was largely bled. Next day the above symptoms were notably exaggerated, and dullness was added on percussion on the right side, posteriorly and laterally. The respiration was puerile over the anterior superior half of the right, and over the whole of the left lung. The bleeding was repeated, digitalis ordered, and nothing allowed in the way of food but milk and cold tea. On the 29th the dullness had invaded

the inferior and lateral aspect of the left lung. The dyspnœa became very urgent, and was not relieved by any treatment, depletory or otherwise ; and he died on the 30th.
Fluid blood, seemingly the product of oozing, was found in
both pleural cavities, and some air also existed on the right
side. Both lungs were much diminished in volume, and
floated toward the upper part of the cavities. The ball
had passed through the second rib, near the posterior superior angle of the scapula, and perforated the apex of the left
lung with a transit of one and a half inches. It had there
pierced the body of the second dorsal vertebra, fracturing
and partially displacing forward its anterior half. It had
then entered the right pleural cavity, traversed the apex of
the right lung, struck and fractured the second rib on the
right side about its center, and finally fell spent within the
pleural cavity. The lungs were gorged with blood, and
their outer and inferior surfaces were coated with lymph.
If one lung only had been wounded, the ball and the effusion might have been both got rid of by operation ; but
when both lungs were implicated, such interferences would
only have hastened death.

The emphysema which was present in this case was probably due to the oblique direction of the wound. It was a
very rare occurrence in the chest wounds which I had an
opportunity of witnessing.

The inflammation which follows gunshot wounds of the
lungs requires the same treatment as that which is given to
inflammation from any other cause. When only a small part
of the lung has been penetrated, then the pneumonia may
be at first localized ; but it will soon spread if not promptly
subdued. During convalescence, the great point which demands attention is to guard against all sources of relapse, as
inflammation is very apt to be re-established, and if it does
reappear, the danger of its giving rise to purulent effusion is
very considerable. Serous effusions often cause much annoyance in cases of wounds of the chest. According to Guthrie,

such effusions take place, in general, from the third to the ninth day, and, if large, imperatively demand early evacuation. I fear this rule was not always attended to during the late war. It is difficult to know what is the best period of the disease to put it in practice.

The strictest regimen should be maintained for ten days or a fortnight after the infliction of a gunshot wound of the lung. Any irregularity in diet, or indulgence in ardent spirits during convalescence, is most apt to cause dangerous if not fatal relapses. Not a few were lost in the East from such carelessness. Opium is of much use in allaying the troublesome cough, which often continues for a long time. Hennen speaks of "a sense of stricture and considerable pain in raising the body to an erect posture, with great anxiety on walking up an ascent," as being frequent consequences of gunshot wounds of the chest; and at another place he says, "diseases which, although we cannot call them pulmonary consumption, agree with it in many points, particularly in cough, emaciation, debility, and hectic, are often the consequences." Veritable phthisis has, however, as is well known, been cured by the rough medication of a gunshot wound. We had no opportunity of watching the remote results of these wounds, as the patients passed from under our care too soon for their development.

Of wounds perforating both sides of the chest, I met with four examples only. In all these the wound was inflicted by grape, and all died in a very short time.

Balls are well known occasionally to become sacculated in the lung. This circumstance, as well as the very small amount of irritation which the presence of such a body may give rise to, was illustrated in the following case. The **case** was first related to me by my friend Deputy Inspector-General Gordon, C.B., and I afterward found the particulars of the early symptoms in the medical reports of the regiments serving in India: A soldier of the 53d, serving in the Punjab, received a ball on the left side of the thyroid cartilage,

which coursed round the neck, entered the apex of the right
lung, traversed it to near its base, and lodged. Violent
dyspnœa, urgent cough, and bloody sputa followed. The
patient, from the fear of suffocation, could not lie down for
several days. These symptoms were allayed by treatment,
and in two months the man was discharged, feeling no inconvenience from his wound. This patient died six months
afterward of a contagious fever, when the ball was found
closely sacculated in the lower lobe of the lung, at the apex
of which a small puckering was seen, but no trace could be
discovered of the ball's track from the apex to its place of
sacculation. The lung was free of disease. In the following
case the position of the ball was not discovered: A soldier of
the Buffs, wounded on the 8th September, received a ball on
a level with but slightly external to his right nipple. Profuse hæmoptysis, fainting, great dyspnœa, oozing of blood
from the wound, and the escape of air followed. He was
largely bled, and his symptoms thereby relieved. Ten hours
afterward, a return of the difficulty of breathing called
for further depletion, and the use of antimony. Pneumonia followed, which implicated the lower half of the
wounded lung. The treatment was that for pneumonia
generally. The wound suppurated, and ultimately closed.
When the patient left the hospital, in December, the lung
acted well throughout, except for a short distance round the
wound, where it was dull on percussion, and seemingly impervious to air. The vocal resonance was notably increased
over the upper part of the wounded side of the thorax.

The direction taken by the ball, and its position as found
after death, give interest to the following case: At the Alma
a soldier was struck by a musket-ball, on the outer side of
the left shoulder. His arm was by his side at the moment
he was wounded. It was observed that the ball had passed
through the head of the humerus, but its ultimate position
could not be ascertained. Nothing was done for the arm.
The ball was supposed to have made a clean hole through

' the bone. A severe attack of pleurisy followed, and on the
subsidence of this, pus was found to point both below the
clavicle and in the axilla of the wounded side. Much bone
came away. Pus flowed copiously by the openings which
were made in the axilla and below the clavicles. The patient became hectic, and died. It was then found that the
ball, having passed through the head of the humerus and the
glenoid cavity, had entered the chest between two of the
ribs, and having run forward within the cavity, and between
the walls and the pleura, had lodged in the anterior mediastinum, where it was found coated with lymph. The chest
symptoms, the surgeon in charge informed me, had been very
slight, and the presence of the ball had given rise to no uneasiness. If the joint, which was the main source of irritation and hectic, had been excised early, a more favorable
result might have followed.

The four following cases are further illustrations of most
severe gunshot wounds implicating the lung :—

At the Alma a soldier was struck by a ball near the center of the left axilla. The bullet escaped on the same level
as that at which it had entered, and within an inch and a
half of the spine. Profuse hemorrhage by the wound and
by the mouth followed immediately and caused the patient
to faint. He was bled at night, as well as next morning, to
relieve the dyspnœa, which was urgent. A severe attack of
pneumonia followed, which, though subdued, recurred on
two subsequent occasions. By December the lung had recovered, except at its base, where it was impervious to air.
The respiration at the summit was exaggerated. There was
in the hospital at the same time another man, whose wound
and its results were exactly similar, only that the ball had
entered by the right axilla in place of the left, and had
escaped a very little lower than in the last case. In this
case the liver escaped injury.

A sergeant was struck at the Alma by a musket-ball, on
the right side, between the sixth and seventh ribs, close to

their angles. The ball traversed the lung, and escaped close
above the inner angle of the clavicle of the same side. The
man said that, on the receipt of the wound, his mouth filled
with blood, and that he fell down and thought he was killed.
Profuse hæmoptysis continued for some days after his admission into hospital. He was largely bled a few hours after
being wounded, and also on the two succeeding days, when
the difficulty of breathing, from which he suffered, became
severe. Tartar emetic was given him and he was kept exceedingly low for several days. Both wounds suppurated
freely. Amphoric breathing was very evident over the
upper part of the wounded lung; but there was no marked
change on percussion anywhere for a week after the receipt
of the injury. He complained of severe pain in the injured
lung during the whole period he continued in hospital.
Three weeks after being wounded, there was a deficiency in
the respiratory murmur all over the right side, which deficiency was balanced by an increase on the left. Broncho-
phony was marked at the upper part of the right side.
There was dullness now on percussion all over the right
lung, but chiefly at its upper part. The expectoration was
profuse and purulent. Cough severe and painful. Pulse
high and irritable. His gums were sore with mercury, and
blisters had been repeatedly applied to the surface of his
chest. He gradually recovered under the influence of a
generous diet; and when he went to England, about four
months after being wounded, both wounds were closed, the
anterior having cicatrized first. At that period the right
side of his chest was somewhat contracted and flattened.
The respiratory murmur was fair over the upper two-thirds
of the right lung, but faint toward the base. Percussion
gave a normal note, except at a small point just at the apex
and at the base, where the sound was dull. A good deal
of bone had been discharged by the wounds during convalescence.

A French soldier had a Minié ball driven through his

right chest at Inkerman. It entered an inch below the
nipple, between two of the ribs, and escaped behind, exactly
opposite the place of its entrance, within two inches of the
spine, fracturing one rib and chipping another. Severe
hæmoptysis and bleeding from the wound followed. He
was bled frequently and kept very low afterward. Most
violent inflammation set in, and effusion took place into the
pleural cavity. The fluid was not evacuated; but while it
was being absorbed, the wound of entrance having closed,
a most violent and prolonged attack of trismus seized him,
which, for a couple of days, threatened to cause death, but
which ultimately yielded to large doses of opium, without
the spasms becoming general over the body. This patient
perfectly recovered and was sent to France.

A soldier of the Guards was struck at Inkerman by a rifle-
ball, which was fired at a short distance behind him by one
of our own men. It entered below the angle of the right
scapula, and escaped between the fourth and fifth ribs, chipping the upper edge of the latter. The hæmoptysis was
very profuse, and much blood escaped by the wounds. He
sank down exhausted almost immediately on receipt of the
wound, and lost consciousness shortly afterward. He lay
a considerable time, he could not say how long, before he
recovered. When he was received into hospital, blood continued to ooze from his wounds, he spat constantly, and his
breathing was greatly impeded. He was bled twice during
his stay in the Crimea, and when I saw him a month afterward, he had in a great measure recovered. The exit wound
had closed, but that of entrance had taken on a phagedenic
action for some days and was not yet healed. The lung
acted well; he could lie on either side; and, to all appearance, he was in a fair way to a complete recovery.

When no adhesions are formed, by which the ball or other
foreign bodies driven into the thorax are arrested, they generally are found lying on the diaphragm, in the angle formed
by it and the costal walls, and close to the vertebral column.

The track of a ball through a lung has been occasionally found to become fistulous, becoming lined by a membrane, and containing curdy pus. The pulmonary tissue around these tracks becomes indurated, and they may or may not have an orifice to the exterior of the chest. A circumscribed abscess may exist between the ribs and the lung, or be in the lung substance itself, and communicate with this track. The perfect manner in which these collections and the track connected with them are closed off from the lung, and the evil which may arise from the presence of this pus, make it a question, which the facts before me do not enable me to discuss, whether or not it would be advisable to evacuate it by operation, seeing that our modern means of diagnosis would permit of its detection. This evacuation could be accomplished by such a puncture through the parietes as would insure the closure of the wound as soon as the object was effected.

CHAPTER IX.

GUNSHOT WOUNDS OF THE ABDOMEN AND BLADDER.

The returns of the war, after April 1st, 1855, show flesh contusions and wounds (simple and severe) of the abdomen, among the privates, as having occurred 101 times, with a fatal issue in 17 cases. There were 38 penetrating wounds with lesion of viscera, and 36 deaths in consequence; while 65 times the abdomen was perforated, and 60 deaths resulted.* Four cases of rupture of viscera without wound were fatal.

The abdominal cavity, from the want of a bony protection in front, as well as from its large surface, is very liable to severe injury in battle, and there is no cavity in the **body** the injuries of which are more serious or more **often fatal.** The ribs protect the contents of the thorax from contusions, and **wounds from pieces of** shell often fail to injure either the lungs or heart; but when a projectile impinges with any force **on** the abdomen, the effects are seldom limited to its **walls.**

---

* M. Legouest mentions 3 cases of penetrating wounds of the abdomen in the Dolma Batchi hospital, all of which died. Alcock reports 19, only 1 of which recovered. Menière mentions 14 in which the ball penetrated, 2 of them being through the side, and all died; while of 7 others, in which the ball passed through the side only, recovery followed. In the Indian wars I find the record of **38** penetrating or perforating wounds of the abdomen, of wh**om 32 died** and 6 recovered. Colles states that in the sieges **of** Moultan "not one case recovered in which **the** abdomen **was fairly** shot **into** and **the** small intestine wounded." **Sédillot tells us** that in the expedition against Constantine they lost all **those** whose abdomens were penetrated by gunshot.

(227)

It is often difficult to tell what influence a certain wound
will produce when it affects the abdomen. At times an
accident apparently severe is followed by trivial consequences, while the most disastrous results may arise from an
injury which shows little external indication of its severity.

Contusions by round shot are among the most dangerous
injuries to which the abdomen is exposed. The hollow or
the solid viscera, as is well known, may be thus ruptured,
and rapid death follow, without much external sign of so
severe an accident. Every campaign furnishes examples of
this. A contusion may, however, arise from a less ponderous missile than a round shot, and the injury be not so
serious. The state of tension of the wall of the abdomen
at the time of the accident appears to exercise no little
influence on the effects produced. When a man is lying on
the ground, and the muscles are completely relaxed, then the
injury inflicted on the contained viscera may be very severe;
but if the muscles are in action and tense, then the force of
the blow will be somewhat mitigated. At least such is the
only manner in which I could explain several anomalous
cases that fell under my notice.

Vomiting and pain in the abdomen are the signs of injury
to which contusions of the cavity generally give rise; and
if no serious damage has been done, all the treatment those
cases require is such as will ward off peritoneal inflammation,
which may steal on very insidiously. If any internal rupture
has taken place, we can do little to prevent a fatal issue.

Shell wounds of the walls of the abdomen are very commonly followed by extensive sloughing, and the danger of
the morbid action laying bare the intestines, or at any rate
favoring their subsequent protrusion, is considerable. In
one case which fell under my observation, nearly the whole
of the anterior wall of the abdomen was destroyed by the
sloughing caused by a shell wound.

Guthrie seems to think that a greater amount of destruction occurs in the abdominal walls than can be accounted

for by their mere injury, this loss being probably caused by their absorption.

Balls often traverse the abdominal walls for a considerable distance without entering the cavity, and they do this at times by so long a transit as to describe half the circuit of the body. Of this very many cases occurred in the Crimea. The strong aponeurosis which protects the front of the abdomen exercises a great influence in deflecting the ball when it has struck at all obliquely. The track which is thus made requires careful management during cure to get it to close. If it be long, it is good practice to make a counter-opening at its center, in order to prevent the lodgment of pieces of cloth or pus in its interior. This can, however, be necessary only when, neither by syringing nor by the introduction of an elastic bougie, we can get quit of them.

Abscesses among the muscles are not uncommon, although very disagreeable complications of gunshot injuries, and especially of contusions of the abdominal walls. **Severe** pain, vomiting, and other symptoms which may be mistaken for those of internal inflammation, may be thus set up.

If the amount of inflammation caused by contusion or other injury of the abdominal wall be limited, then adhesion will take place between the parietes and the omentum or viscera, and will afford a great safeguard against the effusion of blood or other matters into the cavity. If, however, the parietes in part slough, so that the gut is laid bare or opened, the injury is one of great gravity.

It is sometimes very difficult to say whether a ball has perforated the abdomen or not. The relative position, and even the peculiar characters of the two orifices, will not guarantee a decided opinion. Far less can we say, from the apparent direction of the wound, that any of the viscera have been injured. It is neither allowable nor desirable that we should make such a search as will determine the question; for if the ball be not easily found, we never

"amuse ourselves," as Le Dran expresses it, "by seeking
for it," and the treatment ought to be such as will provide
for all contingencies. In the following case, the ball appeared not only to have perforated both the abdomen and
the chest, but also the diaphragm; yet probably it ran merely
under the integuments, possibly traversing the diaphragm
close to its anterior border, and wounding none of the abdominal or thoracic viscera: A ball struck a French soldier
just above the crest of the ileum, and about four inches from
the spine. It escaped close below the inner end of the
clavicle on the same side. At the time he was struck this
man was on his knees, as he was in the act of rising from
the ground on which he had been lying. He had hiccough
and considerable prostration for three days, and also an
attack of pleurisy, all of which he had recovered from a
fortnight after injury, when I first saw him.

The fatality of penetrating wounds of the belly will
depend much on the point of their infliction. Balls entering the liver, kidneys, or spleen are well known to be
usually mortal, although exceptional cases are not rare.*
Wounds of the great gut are also always recognized as
much less formidable than those which implicate the small.
Thomson saw only two cases of wounds of the small gut,
after Waterloo, in the way of recovery; but Larrey reports
several. Gunshot wounds of the stomach are also exceedingly fatal. Baudens records† a remarkable case of recovery,
although complicated with severe head injuries. The syncope which followed the severe hemorrhage in this case
lasted for ten hours, and doubtless assisted, along with the
empty state of the stomach at the moment of injury, in preventing a fatal issue.

The extraordinary manner in which not only balls, but

* See especially the most remarkable case related by Hennen, at
page 455 of the first edition of his admirable "Observations."
† Observations IV., p. 12, of his "Clinique."

also swords and ramrods, may traverse the abdominal cavity, and yet not wound any viscus, has been often dwelt upon by military surgeons.   The escape of the viscera in the following case, which occurred in India, was most remarkable : A soldier of the 28th Regiment, endeavoring to commit suicide, leant over his musket, and drew the trigger with his toe. The ball passed into the abdomen, on a level but a little to the left of the umbilicus, and escaped through the center of the crest of the left ileum behind.   He died in a month. The intestines were found matted together, and large portions of them were gangrenous, but no perforation of the gut could be discovered.   The surgeon, Dr. Young, adds in his report: "This examination, however, in some particulars unsatisfactory, has at least established the fact that the intestines were not perforated by the ball ; but how they escaped, defies any conjecture I can form on the subject." In another case which occurred at Meanee, the ball was ascertained to have gone fairly through the abdomen, yet not to have injured any of the viscera.   It is impossible, however, to be certain of such a circumstance, unless an after-death examination verify a supposition we are too apt to form.

The just and perfect support afforded by the abdominal viscera to one another, and the manner in which they fill their containing cavity, supply a safeguard against effusion after wounds, which has ever been the astonishment and admiration of observers.   The smaller and less torn the wound in the gut is, the more likely is this favorable result to occur. Littre's celebrated case of the madman has ever served as the type of such wonderful acts of "conservative effort." The pressure, too, favors that adhesion between the **viscera,** which is so potent a preservative against evil.

The following case, reported by **Dr. Taylor** when surgeon of the 80th, affords an example of a gunshot wound injuring the smaller gut, while at the same time it shows the effects of such a wound, and also the state of the parts a considerable

period after the infliction of the injury.   It is taken from
the Records of the Medical Department :—

Private Paul Massy was shot through the abdomen at
Ferozeshah.   Very slight symptoms followed, so that it was
supposed the ball had coursed round the cavity, and had not
penetrated.   He mentioned having passed some blood in his
stools after receiving his wound.   The ball had escaped near
the spine, having entered in front.   He recovered slowly
but perfectly, except that he continued subject to bowel
complaint, and finally died of spasmodic cholera, a considerable time (exact period not specified) after being wounded.
For a year before death he was almost constantly under
treatment for dysentery.   When examined after death, the
following was the condition found.   I give it in Dr. Taylor's
own words: "Cicatrix of a gunshot wound in the left linea
semilunaris, about four inches above the crest of ileum; and
on the same plane posteriorly, another cicatrix an inch to
the left of the spine.   Omentum firmly adherent to the internal surface of anterior cicatrix, and gathered into a fold
or knot at that part.   The intestines were neither there nor
elsewhere morbidly adherent; but the fold of intestine immediately opposite to the cicatrix presented a line of contraction, as if a ligature had been passed tightly round the
gut.   The fold of intestine immediately above presented the
same appearance, and on the first fold, four inches from the
first-noticed contraction, and in a line below the umbilicus,
was another similar appearance.   These three contracted
places were of a darker hue, and more vascular than other
portions of the small intestine; having, however, throughout
an arborescent vascularity, and being in the sodden state
constantly seen in sudden cases of spasmodic cholera.   The
mucous membrane of the small intestine was generally of a
pale-pink color.   No ulceration of the large gut.   Upper
part of the colon attenuated, and contracted *in situ*.   Rectum thickened."*

---

* The preparations made of the above parts were sent to Fort
Pitt.

When a ball merely enters the gut, it may be thrown out by stool. Such a case occurred in the 19th Regiment in the Crimea, and is reported by the surgeon in the *Lancet*, vol. i., 1855.

If a vascular viscus be wounded, or a large blood-vessel opened, then hemorrhage may take place within the abdomen to a very serious and fatal extent. The mutual pressure of the viscera does much to prevent bleeding from the former **source**, and the lax attachment of the arteries in general enables them to escape. If blood be poured out suddenly and in quantity, it will partly escape by the wound, and partly collect at the most dependent part of the abdomen, or in the pelvis. Baudens mentions as a certain sign of a quantity of blood being collected in the pelvis, the incessant **and** insupportable desire to micturate caused by the pressure on the bladder, and which is set up, although there is no urine in the viscus. Besides the immediate danger which proceeds from the loss of blood, such effusions, if in quantity, fail to become absorbed, decompose, set up inflammation, and cause death. The quantity must be small which will insure its absorption. It is therefore a matter of some importance to evacuate such accumulations by reopening the wound, rather than to attempt its removal by operation afterward.

The symptoms of penetrating wounds of the abdomen are those which belong to the accident proper, and those which result from its consequences. The collapse is generally very severe, and this is the case, too, in many instances in which the injury appears at first very superficial and trivial. While, in general, this shock and alarm are indicative of deep and serious lesion, they are often excited by no apparently adequate cause. If some hemorrhage, or the effusion of any of the secretions, as bile, or the contents **of any of the** hollow viscera follow the injury, then **the** collapse **will not** only be severe, but will continue.

The subsequent symptoms of these wounds will partake of two characters—those common to all inflammations of the abdomen, and those arising from the inflammation of the particular organ injured. The inflammation which is so certain to occur in the peritoneum requires very careful watching, as it often sets in very slowly and deceptively. "The consciousness of imperfection induced in the cavity," of which Hunter speaks, makes it peculiarly apt to take on an inflammatory action.

The position and direction of the wound, and the concurrent symptoms referable to the lesion of special organs, will lead us to surmise the injury of this or that viscus. The persistent vomiting, the ejection of blood by the mouth or by stool, or with the urine, the escape of special secretions, as bile, by the wound, the peculiar pain or sensation experienced by the patient, will be our chief indications in determining the part hurt.*

The treatment of simple, non-penetrating wounds requires but little notice—the prevention or subdual of inflammation, and the favoring by position of that conservative adhesion between the viscera and the parietes which is desirable if sloughing should set in, so as to endanger the opening of the cavity.

The management of penetrating wounds is not much more difficult, but the results are very much less satisfactory. When the penetration has been occasioned by a ball, it is not often that we have an opportunity of verifying the fact of visceral lesion. No attempt should be made to follow the ball. The

---

* Hunter says of the blood passed by stool: "If it is from a high part of an intestine, it will be mixed with fœces and of a dark color; if low as the colon, the blood will be less mixed and give the tinge of blood;" and of the character of the feeling, he adds: "The pain or sensation will be more or less acute according to the intestine wounded; more of the sickly pain the higher the intestine, and more of the acute the lower." It would be a matter of some moment that we could rely on this sign. We can seldom, however, distinguish the character of the pain from the patient's statement, and it does not always afford us a true guide when it is recognized.

wound should be lightly covered, the patient placed in such a
position as will relax the abdominal walls, fomentations applied by means of the lightest possible material, opium freely
given by the mouth ; and, if inflammation set in, then leeches
and even general blood-letting may be had recourse to.

"All wounds that enter the belly," says Hunter, "which
have injured some viscus, are to be treated according to the
nature of the wounded part, with its complications, which
will be many ; because the belly contains more parts of very
dissimilar uses than any other cavity of the body, each of
which will produce symptoms peculiar to itself and the nature of the wound." "It cannot be too frequently repeated," says Dr. John Thomson, "that copious blood-
letting, and the use of the antiphlogistic regimen in all its
parts, are the best auxiliaries which the surgeon can employ
in the case of all injuries of the viscera, contained within
the cavity of the abdomen." With us in the East the state
of our patients necessitated a much more cautious use of the
lancet in these and in all other injuries, than is **common.**
Opium, however, was the chief reliance in these lesions, as
it allayed that pain and anxiety which might, without it,
have been interpreted into a call for depletion. The most
extreme abstinence from food is certainly one of the most
important points in treating penetrating wounds of the abdomen. Purgatives by the mouth will do harm only, but
clysters, especially of warm oil, are particularly useful and
agreeable to the patient.

Few cases occur in military practice which demand the
use of the suture to the intestine. Such cases are generally
fatal. To those in which its employment is not distinct**ly**
indicated, Hunter's remark particularly applies : "I should
suppose the very best practice would be to be quiet, and do
nothing except bleeding, which, **in cases of** wounded intestine, is seldom necessary."

Early protrusion of the gut is rare, unless the wound has
been occasioned by a large ball, as a grape-shot. Its care-

ful return is, of course, the rule of practice when it does occur. Guthrie has shown the propriety of leaving protruded omentum to act as a plug in the wound.

It is in wounds of the abdomen that the treatment by "debridement" retains its last footing. The fear of strangulation by the strong fasciæ, or between the muscles, is assigned as the claim it has to adoption in these wounds. But experience, while it has overthrown this cause of anxiety, has shown that a positive evil is occasioned by the practice, in so far as that the abdominal walls are weakened by it, and hernia the more apt to ensue. This step then is abandoned here, as in all other regions, unless an absolute necessity arise for its adoption. In the case of narrow wounds through the deep muscles of the back, by which fæces ooze, but cannot get a free escape, in similar wounds penetrating the bladder, or in cases in which a large amount of blood has been effused into the abdomen, it may be necessary to enlarge the wound, in order to prevent ulterior consequences of more gravity than those which can follow from the step itself.

If a false anus result from a penetrating wound by gunshot, the cure will in most cases take place in time spontaneously. Of this I observed, with much interest, two cases at Constantinople, both of which very quickly got well. A plastic operation at a late date will probably supply what is deficient in the effort of nature.

Where the destruction of soft parts has been considerable, the danger of ventral protrusion will require attention during after-life, and no little trouble is often caused by the irregular action of the viscera, by pains which either wander throughout the cavity or localize themselves at the point wounded. These uneasy sensations are increased by any distention, such as that which follows a full meal, and they continue to distress the patient during digestion. Dupuytren dwells on the effects of that chronic inflammation which may be set up by a contusion of the gut, and which, he says, may bring about a stricture of the intestinal canal, or its cancerous degeneration.

I had fewer cases of penetrating wounds of the abdomen under my notice in the East, than of almost any other serious injury. The following are given as among the most interesting of those of which I have retained notes :—

. Cousins, a private in the 77th foot, aged 18, was admitted **into** the general hospital, under Mr. Rooke, on the 8th of June. When standing in one of the advanced trenches, sideways to the enemy, his right arm being stretched out in front of his hip, he was struck by a round shot or large piece of shell, which completely smashed his right forearm, and fractured the ileum of the same side, causing at the **same time** a lacerated wound of the right iliac region about 5 inches long by 3 broad. The wall of the abdomen, including the peritoneum, was destroyed to the extent mentioned, and a coil of intestine was laid bare. No protrusion took place, nor was the gut seemingly injured. Besides the fracture and destruction of the crest of the ileum, the anterior superior spinous process of that bone was quite detached, and the great trochanter was also fractured. The leg on the wounded side was shortened very considerably, and the foot was everted. As, from the extent of the injury sustained and the collapse present, it was supposed that this patient would die shortly after admission, nothing was done for him beyond simply dressing his wounds and giving him stimulants in small quantities. Next day, however, he had so far rallied that some hopes were entertained for him, but it was not till the second day that he had sufficiently improved to allow of his arm being amputated. This was of course done under chloroform, otherwise **it is** questionable whether the **operation could** have been performed at all, the patient was so much depressed. He had at this time no abdominal uneasiness, and his bladder acted freely. By the attentive administration of mild nourishment and opiates, this patient gradually improved. No tenderness or other untoward symptom appeared in the abdomen. The wounds assumed a sloughy look for some days, and deep cellular inflammation in the upper part of the thigh made incisions necessary. On the fifth day his bowels were for the first time moved by the aid of warm-water enemata. At this time the wounds were granulating kindly, and the stump was healing well. The coil of intestine was still visible at that date. The ala of the ileum, which had been laid bare, granulated over, but most of the crest became loose, and was removed at different times. The bowels came to act naturally, and without any stimulation, and by the end of July the wound on the abdomen had completely healed by granulation. The femur, if fractured—and of this there was every symptom, though the state of the pelvis prevented a careful examination being made—became consolidated, but remained two inches shorter than the other. The simplest dressings, and almost no internal treatment, were followed throughout the progress of the case. This patient never had a bad symptom, but made a most excellent recovery; and when he went to England, in September, all his wounds had healed with the exception of two small sinuses, leading to dead bone, on either side of the great trochanter. Below Poupart's ligament, and external to the femoral artery, a hard mass was traceable by the touch, which appeared to be some part of the pelvis driven down into that situation. It did not give him any annoyance. The limb, though shortened, was fully movable at the hip-joint, without causing pain, and he could raise his knee, but not his heel, from the bed. The shape of the hip was destroyed, the projection of the crest of the ileum gone, but that of the great trochanter was unnaturally increased.

O'Neil, private in the 38th Regiment, was admitted, under my charge, into the general hospital in June. A ball entered his left lumbar region, about three inches from the spine, as he was lying on the ground in one of the ad-

vanced trenches, with his feet toward the enemy's works.
The ball lodged. The finger went deeply inward and somewhat upward, but detected nothing of the ball, the situation
of which could by no means be made out. In the evening
his abdomen became a little tender, his pulse hard, and his
face flushed. He was once bled, opium administered freely,
and a fomentation applied to the belly. Next day the
uneasiness had gone, and for eight days there was no return
of it whatever. His alvine evacuations were, in the mean
time, regulated by the use of mild clysters. No blood
appeared by stool. The wound suppurated healthily. **He**
was kept on very mild and easily-digested diet. On the
eighth day severe pain suddenly set up in the left iliac
region. This pain was increased by pressure, but was very
limited in its extent. He vomited frequently, and his pulse
rose to 110 per minute. His bowels had acted freely the
day before. His tongue was dry and furred. He had **a**
dozen leeches and repeated fomentations applied **to the**
abdomen. Dover's powder, in doses of gr. x, was ordered
every second hour. Next day the pain had quite left, and
all treatment was stopped. His bowels did not act without
the use of a clyster. He got plenty of mild nourishment,
and, after a time, cod-liver oil. Though without any
uneasiness or symptom of ailment, he became much emaciated, but ultimately rallied, and made a good recovery, the
position of the ball never having been discovered, though
the direction and depth of the wound would appear to
favor the view that it had penetrated the cavity.

I saw a patient in one of the French hospitals at Constantinople whose abdomen had been traversed from behind,
forward, by a ball at Inkerman. The bullet had entered
near the spine of the last dorsal **vertebra, and had** escaped
near to but slightly to the **left of the** umbilicus. The gut
protruded for some days at the anterior wound, but did not
appear to be injured, at least no intestinal secretion showed
itself at either orifice. Hardly any bad symptoms seemed

to have followed. The gut was returned, the man kept low, and opium freely administered. He made a most excellent recovery. In another patient in the same hospital, a wound of exactly the same description had been inflicted. The same symptoms and result followed, except that the gut did not protrude, and that recovery was slower.

The following was a very remarkable case, which, though not strictly a wound of the abdomen, I mention here, as I do not intend to refer to gunshot wounds of the rectum. I saw the patient at Scutari, toward the end of 1854, under the immediate charge of Mr. Price, now assistant-surgeon of the 14th Regiment. A ball entered the front of a soldier's left thigh, three inches above the patella, as he was mounting the heights at Alma, and passed upward deep among the muscles of the thigh. It then turned round the limb, traversed the muscles of the left hip, crossed the perineum deeply, and escaped on the right hip, having passed through the rectum some way above the anus. The wound of exit closed, and for several days before death fæces passed by the wound above the knee. Sloughing and irritative fever set in, and he sank rapidly.

To prevent the infiltration of fæcal matter in these cases, Larrey has recommended the use of a tube in the rectum.

The bladder has been wounded by gunshot several times during the past war, but the returns fail to tell us how often.

Balls at times pass through the pelvis, and yet spare the contents.* Thus in one case, of which I have notes, it passed in by one sacro-ischiatic notch, and out by the other, without doing more mischief than contusing the rectum. When the bones of the pelvis are broken, the injury is very serious, from their deep position, neighborhood to

---

* In the case of a man wounded at Chillianwallah, a six-pound grape-shot passed through the pelvis, and yet he survived four days.

important vessels, and thick covering. Stromeyer has called attention to the great liability there is to pyæmia after such injuries. If the ball passes through the peritoneum, then the risk of violent inflammation is so great as to render the wound generally fatal.

The bladder may be wounded in many directions; but the passage of the ball in an oblique line from above downward, and to either side, seems the most common course for it to take. Occasionally its superior fundus is opened by a ball passing across the abdomen from side to side, close above the symphysis pubis. The gravity of the wound will depend mainly on whether the peritoneum has been injured or not. If it has not been opened, then the prognosis will, in some measure, hang upon the empty or full condition of the viscus at the moment of penetration. If the direction of the wound permit of the infiltration of urine into the peritoneum, then the fatal issue will not be long delayed. These are the cases whose hopeless nature probably gave rise to the oft-quoted Hippocratic axiom, " Cui persecta vesica lethale ;" as gunshot wounds, at any rate, implicating those parts of the viscus which are uncovered by serous membrane, are by no means so mortal as they were so long supposed. Dr. John Thomson saw in Belgium alone fourteen cases in a fair way of recovery.

A ball may lodge either in the neighborhood of the bladder, or, entering its cavity, remain there. This latter result will be most apt to occur when the bladder is full of urine or the ball much spent at the moment of contact. In rare cases, a ball, when very small, has been passed with the urine, and it has been known to escape by the formation and opening of an abscess in the perineum.

The urine may escape by the wound at once, or at a later period when the eschar separates from the wound, or it may not escape at all. It is seldom, however, **that it** fails to pass in some quantity at the time of injury. The swelling which takes place in the lips of the wound prevents in a great measure the flow of the secretion by the opening; but it is by no means always sufficient to do so, as we would be led to suppose from Larrey's statement. The urine may, and does at times, escape by both wounds, if the ball has passed out; but from the greater amount of bruising and swelling which takes place at that of entrance, it may fail to appear there, even although it be the more dependent, and flow only from the wound of exit. The early passing and retaining of an elastic catheter is a most important part of the treatment of these cases, as it prevents the urine, in traversing the canal of the wound, from becoming infiltrated among the divided tissues. Larrey, recognizing the existence of this danger only at the period of separation of the eschars, did not employ a catheter early, but was particular in its use at the period when he thought the accident referred to was most apt to occur. Moreover, the fact that the slough is by no means the barrier to infiltration **which he supposed it** to be is now well recognized, as well as that the exact period when its separation is to be looked for, we know, cannot be relied on. The irritation and straining which the unevacuated urine occasions may prematurely force off the slough, and allow the urine to become effused, and so the mischief may be done before **we are ready** to combat it. Unless the wound implicate the neck of the bladder, the presence of a gum catheter will create but little irritation, and should be enjoined from the moment of injury. The catheter had best be retained till the urine begins to flow by its side, as the formation of abscesses, with their disagreeable and dangerous consequences, is thus more safely guarded against.

Larrey, with the object of obviating infiltration and venous engorgement, had recourse to scarifications, so as to enlarge the wound and prevent all retention of secretion in its track. This step will, however, be perfectly uncalled for, if the catheter be retained from an early period. Rest, low diet, mucilaginous drinks, enemata, it may be leeches, and fomentations, or hip baths, will comprise the rest of the treatment in the majority of cases.   The employment of morphia suppositories will also be found, under certain circumstances, most useful.   If any urine does escape into the tissues, its early evacuation will of course be necessary.

The posterior or lower wound commonly closes before the anterior; but neither ought to remain long open, if the catheter be made to remove the urine so soon as it enters the bladder.   If the part through which the ball has passed be deep, the external orifice of the wound may close before the rest of the track—a result which should be avoided.

The position of the bladder, its depth from the surface, its size internally, the want of correspondence which takes place between the external wound and that in its walls from their contraction after the passage of the ball, make the extraction of a ball by the wound a matter of impossibility without such an enlargement of the orifice as would be injurious.

If the ball remains in the bladder, it becomes a matter of moment to remove it.   Balls, pieces of cloth or **bone, so** introduced, form the nucleus **of** calculi; so that the sooner they are got quit of the better, provided the immediate irritation and inflammation caused by the wound have subsided. Many cases **are now on** record in which the bladder has been opened, and calculi, having balls as their nuclei, have been removed.   Larrey operated successfully on the fourth day after the introduction of the ball, and mentions a case in which Langenbeck succeeded in removing a similar body ten years after its introduction into the bladder.   Morand operated twice.   Demarquay mentions a case in which the nucleus was a piece of shell.   Baudens successfully removed the ball by an incision above the pubis; **Guthrie, by the lat-**eral operation.   Hutin mentions two cases **in which** a ball or foreign body was removed by lateral incision—one after thirty-two years' and the other after nineteen years' residence **in the** bladder.   In one of these cases, three calculi were removed, having pieces of cloth as their nuclei. Besides these, Mr. Dixon, in the 33d volume of the *Medico-Chirurgical Transactions*, has given the particulars of ten other cases in which balls were successfully removed, and three in which the attempt failed. Nearly all of these patients were operated on years after being wounded. In the *Medical Examiner* for 1855 a case is recorded in which a large ball, driven into the bladder, was not found till two years after, on the death of the patient. It formed the center of a large calculus concretion.

The following case I find detailed in the Report from the sanitary depot at Landour for 1849–50 :* Private West was wounded on the hip by a grape-shot at Chillianwallah. The ball was lost, and the wound healed kindly in six weeks. A day or two after being wounded, he experienced a scalding sensation in the urethra on micturating, and he showed marks of a urethral discharge on his linen, which he thought was a return of an old gonorrhœa. He was treated under this idea for a time, the symptoms of inflammation in the bladder being ascribed to the gonorrhœa. The attacks of cystitis became so severe as to cause his bladder to be examined, when a hard substance was discovered. The introduction of the instrument gave great pain, and it was only on the second trial that a foreign body was detected. By the lateral operation a grape-shot was found and extracted, "slightly incrusted with a sandy deposit." He recovered perfectly. No bone was injured by the ball. "After the operation, the patient remembered that he used to pass blood and pus in his fæces after he was wounded. Hence it is probable that the ball entered by the sciatic notch, and traversed the rectum, entering the bladder at its back part."

---

* Unpublished records of Medical Department. This case is referred to by Guthrie, and has been recorded by Mr. M'Pherson, in connection with Mr. Dixon's paper, but with some variation from the account given in the text.

The following is a fair example of a penetrating wound of the bladder :—

Griffith, private 57th Regiment, was admitted into the general hospital in the summer of 1855. A ball had entered his left hip, close to the tuber ischii, and escaped on the abdomen, two inches above the symphysis, a little to the right of the middle line. Urine escaped by the anterior opening. A catheter was passed into the bladder and retained there. He had no bad symptoms of any kind for twelve days. His urine passed by the catheter, and also by the opening on the abdomen. His pulse remained quiet, and his abdomen without uneasiness. His general health was unimpaired, and his bowels acted regularly. The posterior wound, through which urine never passed, closed rapidly. On the twelfth day he had severe pain in the abdomen, which was, however, relieved by a dose of opium, and he never afterward had a bad symptom or uneasy feeling, except the irritation occasioned by the urine flowing on the abdomen, which could not be altogether prevented. His urine was loaded with mucus and pus during the period of cure, and he passed several small pieces of bone, both by the urethra and by the abdominal wound. At the end of six weeks he could retain his urine, and pass it at pleasure by the natural passage in a full stream. For a month he had been unable to prevent his urine flowing constantly away. In about two months from the period of his admission the wound on the abdomen was completely closed by the use of nitrate of silver. His strength, which had somewhat failed, was at that time quite restored, and he was walking about the ward convalescent. At this period he passed from under my notice; but I learned that the wound on the abdomen had reopened, and that he could pass his urine without any pain through this opening in a continuous stream, but that ultimately, before he went to England, it had permanently closed.

The following case is curious, as showing how large a body

may descend into the pelvis, and yet very slightly injure the
viscera: A soldier at the Alma was wounded by a piece of
shell, which struck him over the symphysis pubis, and, descending into the pelvis, was lost.   No bad symptom whatever supervened, and he made a rapid recovery.   The surgeon in charge of the case thought that the missile lay
impacted deep in the pelvis, behind the pubes, but this he
could not satisfactorily determine.   Here the bladder escaped
most miraculously.

The injury was much more severe, but the result little less
fortunate, in the following case :   A French soldier of the
line was struck at the Alma by a piece of shell, above the
symphysis pubis, which fractured the bones, passed downward, and was removed in the perineum from the side of
the urethra.   The rectum and urethra were both lacerated.
Deep abscesses formed, the patient's strength gave way, but
no acute attack of inflammation seized any of the viscera.
A communication was established between the bladder and
rectum, and between the bladder and the abdominal **wall, so**
that gas and small pieces of fæces escaped at times on the
abdomen.   Blood frequently passed by the urethra.   The
last time I saw this man was in January, 1855, when he was
recovering rapidly.

In the next case the missile penetrated the pelvis from
below, and it is interesting chiefly from the manner in
which the peritoneum escaped.   A French artilleryman was
wounded at the battle of the Alma by a piece of shell, which
struck **him** on the perineum, and penetrated between the
rectum and bladder, establishing a fistulous communication
between these parts.   The peritoneum was not opened.   No
bad symptom followed, but when he was sent home he was
dying of phthisis.

There is a case related in one of the Indian reports, which
illustrates in a curious way the severe injury which the perineum may undergo.   A soldier of the 14th Light Dragoons

had the pommel of his saddle struck by a round shot at Goojerat. The ball passed under and between him and his horse, which escaped injury. The rami of the ischium **and** pubes were fractured on the left side, the perineum extensively lacerated, but the scrotum was only slightly abraded, and the urethra was uninjured. He had much pain afterward in passing his urine; the soft parts of the perineum sloughed, and his testicles atrophied, but otherwise he made a good recovery.

CHAPTER X.

In the returns of the late war, from April 1st, 1855, 2198 cases of gunshot
wounds of the lower extremities appear among the men, and 166 deaths
therefrom. Of these, 1628 cases and 55 deaths were mere flesh wounds,
and 43 cases and 2 deaths wounds with contusion and partial fracture
of long bones; 23 cases and 1 death, simple fracture of long bones by
contusion of round shot; 174 cases and 64 deaths from compound fracture of the femur; 66 cases and 9 deaths from the same injury of the
tibia or fibula alone; 144 cases and 27 deaths from compound fracture
of both bones of the leg; 88 cases and 7 deaths from perforating or
penetrating wounds of the tarsus. Besides those who died directly from
the injury, 96 cases of compound fracture of the femur, and 91 cases of
compound fracture of both bones of the leg, were submitted to amputation.

There were 1237 cases and 8 deaths from flesh wounds of the upper extremity; 102 cases and 12 deaths from contusion and partial fracture
of the long bones, (including the clavicle and scapula;) 27 cases and 2
deaths from round shot simple fractures; 169 cases, 15 deaths, and 104
submitted to amputation, from compound fraction of the humerus; 66
cases, 2 deaths, and 41 amputations from compound fracture of the bones
of the forearm. In 113 cases the structures of the carpus were penetrated or perforated, and 48 of these cases were subjected to amputation.

OF all the severe injuries received in battle, none are of
more frequent occurrence or of more serious consequence
than compound fractures. They cause peculiar anxiety to
the surgeon, from the manner in which their extent and
gravity are so often masked, and from the uncertainty which
still prevails as to many points in their treatment. This
ambiguity as to their management arises in a great measure
from the many varying causes connected with the state of

(248)

health of the patient, and the means at hand for his treatment—circumstances which fluctuate with every campaign.

In the Crimea, these injuries were peculiarly embarrassing and extraordinarily fatal.   In the management of no accidents was so much expected from modern improvements, and by **none** were we so much disappointed in the results. It was confidently hoped that in very many of those cases which, in the old wars, would have been condemned to amputation, the limb would now be preserved, either by the **exercise of** greater care in the treatment, or by having recourse to some of the modern expedients by which limbs are so often saved at home.   But, unfortunately, a sad experience only confirmed the hopeless nature of compound fractures of the thigh by gunshot, and their very uncertain and dangerous character when the leg or arm are implicated.

In the following remarks on compound fracture, I propose to refer chiefly to those cases in which the femur was broken, and I will notice afterward similar **injuries of the leg and arm**.

It can **hardly be doubted that · the great** striving after conservatism, **which** influenced all the surgeons of **our** army, was one main cause **of** that mortality which attended these injuries.   **We** were not prepared to believe how hopeless they were, till the unwelcome truth was forced upon **us** by an ever-recurring experience.*   We were disposed to judge

---

* "When the brigade was first landed, an opinion prevailed that **cases of** compound fracture of the thigh would be met with in which **it would be** proper to attempt to save the limb, and every case was carefully examined in order to determine the kind of treatment it would require.   Two cases were at length brought into the camp, respecting which the majority of the medical officers were of opinion that amputation should not be performed.   The men were both young, healthy, and temperate; the injury to **the bone and** soft parts comparatively slight in both **cases**.   The **external** wound was small, and situated on the outer side of the limb **in** one, two, or three inches below the trochanter; in the other, the same distance above the knee.   **The** result of the former was, after great suffering,

of compound fractures by gunshot as we would of accidents, similar at least in name, seen in civil life. **Full of the prom-**ise of the schools, we would not admit that any injury appa-rently so slight could withstand the assiduities of a wise conservatism.   In trying, however, to save limbs we lost many lives, thus fulfilling the prophecy of one **of the** greatest surgeons.* Cases of promising appearance were reserved for the trial — the very cases, in fact, which would have made the best recoveries if operated upon early, and

---

death.   **In the** latter, although the patient recovered, there is little prospect of the limb ever becoming useful. Experience has therefore forced upon us the conviction that to attempt to save the limb in any **case** of compound fracture of the thigh, the result of gunshot, is to endanger the patient's life; and the result of secondary amputation **has not** been such **as to** induce us to trust to that chance of saving **life after the** failure **of** the first attempt."—*Report of Black Sea Fleet,* p. 36.

* How similar has been the slow process of **conviction** on this head, is well shown in the following **remarks of M. Gaultier de Claubry,** quoted by Paillard in a note: "Lorsque j'arrivai sur le théâtre de la chirurgie militaire, je me permis de blâmer hautement la conduite **de** mes chefs, que j'appelais aussi *routinière, barbare;* je parvins même, à force d'instances, à force d'assurance des resources de la nature et de l'utile secours de l'art, à porter quelques chirurgiens militaires a douter de la justesse de leurs déterminations; à hésiter, dans certains cas, à s'armer de l'instrument tranchant. Eh bien ! les plus expéri-mentés m'assauraient que je ne tarderais pas à revenir de mon erreur; les autres ne tardèrent point à gémir avec moi, eux, de leur blâmable condescendance, et moi, de la présomptueuse légèreté avec laquelle **j'avais jugé, une** conduite sanctionée par une longue expérience, sans avoir **réuni tous les éléments de** la question. J'ai encore présents à l'esprit les nombreux blessés de la compagne de 1805, en Italie, chez lesquels je passais des journées entières à panser des fractures com-minutives des os longs, et qui succombèrent tous, **les** uns dans les premières jours, par l'effet des accidents primitifs, douleurs, convul-sions, fièvre, résorption purulente; les autres après un temps quel-quefois fort long, lorsque leurs blessures avaient éprouvé un notable amendemente, par l'effet du typhus nosocomial, de la dysenterie épidemiqué," etc.

the inevitable amputation was delayed till the patient's constitution had become so depressed as to be beyond reaction.

Two circumstances seem to have had chiefly to do with
the irreparable character and mortality of compound fractures of the thigh in the Crimea—first, the state of health
of the men when wounded; and, secondly, the effect on
bone of the new kind of ball with which most of these injuries **were** inflicted.

As to the state of health of our patients, it was not
merely that they were in so anemic a condition that suppuration and irritation quickly prostrated them; nor was it
that their stamina and "pluck" had been destroyed by hardship and suffering; nor that the means of treating them in
front during the early period of the war were totally want-
ing; but the chief cause of the reluctance shown by nature
to repair the osseous breach was the scurvy-poison which
held command in their systems. This it was which mainly
opposed recovery. Callus was not thrown **out at** all; or **if**
it was, it refused to **consolidate.** I myself examined **the**
limbs of a large number of men who died at Scutari during
the early part of the **war,** and in not a single instance
almost did I observe **the** slightest attempts at repair; but,
on the contrary, invariably found a large sloughing chamber
filled with dead and detached fragments of bone, shreds of
sloughing muscle and destroyed tissue into which the black
and lifeless bones projected their irregular extremities, and
across which, lying in every direction, but seldom in the axis
of the limb, were dead and detached sequestra, the "frac-
ture-splinters" of the accident.

The depressed condition of body to which the hardships
of the war had reduced the men made a severe compound
fracture of the femur synonymous with death; so that we
might with perfect appropriateness use **the** words of Ravanton : "I exhausted many times the resources of art without
success—incisions, removal of the fragments, early bleed-
ings of sufficient magnitude, spare diet, dressings, position,

infinite care, nothing could protect them against an inevitable death." Most of our patients, as I before remarked, had either suffered from dysentery or were on the verge of falling into that disease. The vast majority of them had ulcerated intestines, and were thus in a condition of health which did not bear disease. When men in this state received a severe compound fracture, and their constitutions were taxed to repair the injury, there was no reserved fund on which to draw. They had been living up to their income of health, and so utter failure was the sure result of increased expenditure. If when injured they had been taken into the ward of a London hospital, I doubt whether they would in most cases have ended more fortunately, either by preserving the limb or by amputation : how much less, then, when they had to undergo treatment in a camp!

Many of our patients looked very well at first—appeared, perhaps, strong enough, and expressed such a confident hope in the result as almost to deceive their surgeon. The injury might not appear very severe; the bone was undoubtedly broken, but it might not be much comminuted; and thus we flattered ourselves, and began a trial hopefully which always ended in disappointment. The golden opportunity was allowed to pass, and so we entered on a road which led to death, whether through the portal of amputation or any other. The struggle soon began. Suppuration set in. The disease which lurked in "blood and bone" showed itself. Diarrhœa appeared and would not cease. The patient's stomach refused the only food which could be procured. He got emaciated, weak, and irritable. A suspicion was awakened that the bone had been more severely injured than was at first supposed. Things went on from bad to worse. Hectic claimed its share of the waning strength; and whether we operated late or not, the great regret remained that it was not done at first, as the invariable result demonstrated the uselessness of any other proceeding.

During the greater part of the siege, the means of treat-

ing these accidents, whether as regards food, bedding, clothes
or shelter, did not exist in camp; and to transfer them to
the rear only made the fatal result the more certain, from
the pyæmic poisoning which was sure to be set up by the
transport.   Thus, then, it came to be that up to the period
when things were improved in the camp hospitals and in the
transport service, recovery from a compound fracture of the
thigh was impossible, or nearly so, and that the best hope
lay in an early amputation.   The only exception to this I
will afterward allude to.

Now, while it may with seeming truth be said that, as
most of these circumstances were peculiar to the Crimean
war, the principles deducible from them are not of universal
application, still many of them are inseparably connected
with warfare.

It is essential to the successful treatment of compound
fractures, that the patient be supplied with suitable food,
that his broken limb should be retained for a certain time·
immovably fixed in a proper apparatus, and that it be carefully and regularly dressed.   But how can these things be
guaranteed in war?   In a siege they ought to be more feas-
ible than in any other kind of campaign; but how was it
with us?*   Besides the privations which most armies
undergo, there is the inevitable shifting of position, and
of the wounded, unless the plan Esmarch tells us they

---

* At the siege of Antwerp, where every convenience existed, and
the men in good health, their experience was no more favorable than
usual.   In the Sleswick-Holstein war they saved a few—the hardships being very greatly less than those which can be looked for in
any other war.   In India, also, a good many compound fractures
were cured; but then the difference between the match-lock ball and
the conical is very great.   The Sikhs used a ball which weighed only
3 drachms and 40 grains Troy, and had very much less propulsive
force than ours, as is evidenced by the number of cases mentioned
in the Reports, in which they failed to penetrate, or did so only
superficially.   The Chinese match-lock was also a weak weapon.

adopted in Holstein be followed, and the patients left to fall
into the hands of the enemy in order to obviate the necessity
of transport; but I fear there are few medical services which
could afford to do as he proposes, viz., leave a certain number of their body to be taken prisoners, in order to secure
the unremitting and careful treatment of the cases.*

So much for the health of our men; let me now refer to
the second cause, which, I believe, rendered our fractures so
unmanageable. There cannot be a doubt that the old round
ball, if fired at a certain range, comminuted bone, but it is
equally certain that at a longer or shorter distance, it frequently failed to do so.† When fired a few hundred **yards**
off, it had hardly force to enter the body, but might be
diverted, as it has been, by the point of the nose. If it did
enter, and impinge upon bone, it might only dent it, as may
be seen by an example in the Musée Dupuytren, in Paris;
or it might groove it merely, or, penetrating the substance of
.the bone, it might remain at rest without splitting it, as can
be verified in any museum of a military hospital. If fired
again, at close range, the round ball might go through a
bone, making a bore as clean and sharp as if formed by a

---

* "Thus we foresee," says John Bell, "an argument of necessity
as well as of choice, and that limbs which in happier circumstances
might have been preserved, must often, in a flying army or in a dangerous camp, be cut off. It is less dreadful to be dragged along with
a neat amputated **stump than with** a swollen and fractured limb,
where the arteries are in constant danger from the splintered bones;
and where, by the least rude touch of a splinter against some great
artery, the patient in a very moment loses his life."

† **I believe** that the proportion of cases in which **balls** have passed
through the fleshy parts of an extremity **without** fracturing the bone,
will be found to be much **less in** the Crimean than **in** other wars.
Thus, in one series **of cases** mentioned by **Deputy-Inspector** Franklin
in his report on the wounded at Meanee, 31 cases of match-lock
wounds of the upper arm are given, and in only one was the bone
broken. To work out this point in figures so extensive as to be of
any use, would require details not supplied by writers on the old
wars.

punch. Of this fact many illustrations can be seen in surgical museums. Now, so far as my observation goes, none of these results follow the stroke of a heavy conical ball, such as that used by the Russians, at whatever range it is fired. It never rests in a bone, channels or perforates its substance, without splitting it, like a wedge; nor does it ever come to mark a bone with any touch more gentle than what occasions its utter destruction. In the Crimea we had many opportunities of observing the action of both kinds of ball, and, so far as I could judge, their effects were so dissimilar as almost to justify a classification of injuries founded on the kind of ball giving rise to them.* The longitudinal splitting of the bone is so dextrously and extensively accomplished by these balls that, while but a small opening may lead to the seat of fracture, the whole shaft may be rent from end to end. I have repeatedly seen the greater part of the femur so split. Stromeyer has shown that this longitudinal splitting seldom transgressed the line of the epiphysis, an observation which I can most decidedly confirm; **for** though the injury has at times been sufficiently severe to implicate both, yet the rule has been just as he says.

Gunshot fractures of the **long bones** of the extremities have always been considered dangerous, chiefly on account of the shock, the comminution of bone, and the fact that the wound leading **to it** is of such a character **that** it can heal only by suppuration, and cannot be so closed as to convert it into a simple fracture, which, it is well known, we **can** sometimes accomplish in such fractures as present themselves to us in civil practice. The cavity of the fracture is thus kept open to the air; the pus undergoes those changes which Bonnet **has shown it** does **under** such **circumstances,**

---

* In these remarks **I refer merely to the heavy conical ball,** as there are balls of the same shape, **but of less** weight, which are by no means so formidable. That **used** in the Sleswick-Holstein campaigns appears to have been **very** trivial in comparison to the large Russian one, of which we had **such** dire experience.

and that severe and prolonged inflammation of the deep and
irritable tissues, which constitutes the chief danger in compound fractures, cannot be avoided.* Now, all of these
dangerous characteristics of compound fractures have been
immensely increased by the conical ball. First of all, the
shock it occasions is undoubtedly greater than that **caused**
by the round ball, simply because the destruction it causes is
much more severe. Secondly, the comminution of bone is
enormously increased; the number of fragments which are
quite detached are much more numerous, and the amount of
sequestra, which are so far severed as to be ultimately
thrown out before a cure can be looked for, is much greater.
Thirdly, the bruising of the soft parts is more extensive, so
that the suppuration is more prolonged, and the changes of
purulent absorption so much the more multiplied.

The great loss of substance which follows compound fractures by the conical ball is the source of one peculiarity in
their treatment. The shortening will be greater should consolidation follow, than if the injury had been occasioned by
the round ball. The conviction has been strongly impressed
upon my mind, by the observation of not a few of these

---

* "All the complete fractures of the other bones of the extremities unite when they are well managed; by what fatality are those
of the femur not equally fortunate? Is it the diameter of the cavity
of the bone; the quantity of medullary substance which it contains;
the peculiar structure of the vessels which carry the nourishment;
the size and force of the muscles which are attached to it, which, by
their weight **and pressure,** obstruct the passage of the liquids? All
these causes united may combine together, and give rise to that want
of success **which we meet with in treating complete fractures of the**
femur, caused by **firearms; but complete fractures of this bone heal
very** well, whatever cause **has produced them, when they are not ac-**
companied by a wound. These **reflections, which the bad** success of
those cruel fractures has suggested, **have caused me** to present to the
public, in 1750, a method **for** amputating the thigh at the hip, *and
that to try and snatch the wounded from inevitable death.*"—RAVANTON,
Chir. d'Armée, p. 324.

cases, that we ought not to keep up extension in their treatment, except in a very modified degree.   If we do so—if we drag and haul at the bone, as I have often seen done—what is the result?   A large hiatus exists, void of organizable material for forming the bone; the parts active in repair are drawn far apart, and a tax is made on the reparative process, which I will not go the length of denying may, under the most favorable circumstances, be brought about; but which I am fully certain never could be accomplished with us.   In many cases it would, to my mind, be better practice—*i.e.* it would afford better results in saving life and limb—rather to approximate than draw apart the fractured ends in such cases.   Allow the ends of the bones to be drawn by the muscles toward one another, *having first removed the sequestra*, and attend merely to keeping the limb as straight as possible; or, in other words, do not be troubled with the displacement as to the length, but only as to the thickness of the bone, and I believe our chance of success would be improved.   Deformity we would unquestionably have—shortening and twisting, and a limb of **which** I, for one, by no means recommend the keeping; but if we *must* save the extremity, if its retention is to be the test of good management, then I think our hope must be in some step like the foregoing.

There are rare instances of compound fracture, which seldom present themselves now-a-days, in which the bone is but little comminuted, and which demand a different consideration altogether from those I have been speaking of. These accidents commonly arise from the contusion of a round shot, or the contact of a piece of shell.   They are, however, so very rare and difficult to recognize that **less** harm will follow from the same line of practice being pursued with them, viz., that of immediate amputation, than if, by being careful about such rare exceptions, we run the extreme hazard of sacrificing the majority of cases which determine the rule.

22*

The extensive comminution of the bone by a conical ball makes the indications with regard to the management of the sequestra more evident than it is commonly considered. I do not think we paid sufficient attention to their removal in the East. It may be true, as some tell us, that in fractures with the old ball, it was desirable to meddle as little as possible with the fragments; but this is the teaching of only a few. However, to my mind the question assumes a totally different light when viewed by the pathological results we had occasion to witness. It may be remarked, before proceeding further, that it is impossible not to recognize the practical nature of the division of the sequestra made by Dupuytren into primary, secondary, and tertiary, according to their degree of connection with the parts, and this, notwithstanding Esmarch's assertions to the contrary; nor can I see that the distinction of them, proposed by the latter, into "fracture-splinter" and "necrosed-splinter," makes the thing a whit clearer, or the division a bit more useful; so that in the following remarks I will adopt the old division.

The longitudinal sections into which the bone is split are mostly capable of consolidation, except at points where their connection, or the contusion they have undergone, places such parts of them in the position of tertiary sequestra, which will exfoliate at some undetermined date. These fragments cannot of course be touched. The secondary splinters, again, or those loosely connected—hanging by an extremity or by an edge to the periosteum or to the tissues —are commonly very numerous, and lie by their detached parts in all directions to the axis of the shaft. The primary sequestra, or those wholly separated from their connection by the accident, are, in fractures from the conical ball, peculiarly numerous and destructive in their action. In some cases which I have had an opportunity of examining, these were found not only at the seat of fracture placed in every possible position except the right one, but also driven deeply into the soft parts on the side of the limb next the wound of exit—long, sharp, delicate chips, whose presence must have been the cause of continued suppuration, of low, disorganizing inflammatory action in the soft tissues and bone, which extended its ravages to limits **far** beyond the seat of injury. In one case which I observed in camp, where partial consolidation had taken place, the dead sequestra had become so involved in the new bone, and were so prominent, so irregular, and so rough, as to look like the bristles of a porcupine.   When to these considerations we add the chance of other foreign bodies, pieces of accoutrements or cloth, remaining between the broken fragments, and the ideas suggested by the very narrow opening to the surface which remains in gunshot wounds, further reasons will be seen for the practice which, I believe, should be in general followed —namely, enlarging the *exit* wound, (especially if it be the more dependent, or if it be a conical ball which has occasioned it,) extracting all loose and slightly-attached fragments, and keeping the aperture open, so as to allow **of** the free flow of the pus.

We have seen that the severe commotion at the **seat** of fracture occasions the formation of that large "foyer" which is found full of detached and dead sequestra, disorganized tissue, and acrid pus, and which, unless it be got rid of, continues to bathe the ends of the shaft, gives rise to inflammation in the medullary membrane, supplies a depot of absorption for the uncollapsing veins of the bone, and finally causes constitutional poisoning.   Now, as a ball traversing a limb carries the fragments it detaches toward its place of escape, it is evident that they will be the more easily got hold of and removed on that side of the limb.   These are the grounds on which the practice, advocated above, is founded.   Unless such a step as is indicated be had recourse to, I cannot see how it is possible, except in very rare and exceptional cases, to hope for the cure in the field **of a com**pound fracture of a large bone by a conical ball.   Dupuytren, recognizing the necessity of getting quit of these fragments, recommends the enlargement of both orifices to an extent so great as "that the fingers, introduced by either opening, should pass freely and meet without impediment."* This, he thought, however, should be avoided, if the part was very thick and muscular. The proceeding sketched above is in no way so severe as this, and would be probably as efficient in fulfilling the end in view.

All surgeons who have had much to do with gunshot wounds are agreed as to the propriety of removing those fragments which are wholly detached; but some oppose the removal of any which retain the least attachment. The objections which have been advanced against the extraction of these are, chiefly, that they assist in the repair of the breach, by throwing out bone, and that if they do die, they will be extruded by the suppurative process. To this it is replied that, if these fragments are at all extensively attached, their removal is never contemplated, but that if they are connected only by a border or an end to the shaft or the periosteum, they can contribute but very slightly to form callus, and will almost in every instance die. One small part that is covered by periosteum may generate callus, but the rest of their bulk will surely perish, and give rise to abscesses and fistulous openings; and the amount of irritation, constitutional disturbance, and wasting suppuration which they will cause, before they are thrown out by the eliminative force of nature, are such as to make it impossible for any but those whose constitutions are the strongest and most vigorous to withstand it. The length of time during which these spiculæ keep up the suppuration and retain the wounds open not only render the patients the more subject to pyæmic poisoning, but, what is of some consequence in military practice, detain the men longer in hospital; thus encumbering the wards, and keeping the patient longer exposed to

---

* This is a mere repetition of the opinion of Percy, p. 188 of his 'Manual.'

an attack of those fatal forms of gangrene which prevail in
such circumstances.

It is needless to quote authorities to show how practical
experience has condemned the leaving of these secondary
sequestra in the wound, as nearly all military surgeons are
agreed on the necessity for their removal. M. Begin thus
formulizes his great experience in a communication to the
Academy : **"I do not** know any precept more erroneous and
more dangerous in surgery, than that which tells us to
respect and retain the fragments of bone partly detached in
fractures.   These fragments almost never recover their
vitality, nor become united to the body of the bone;" and
he also tells us in another place, to remove not only "those
pieces which are entirely detached, but also all those which are
movable, vacillating, and capable of being extracted without
the necessity of too great destruction." M. Hutin, again,
whose position in the Hôtel des Invalides gives him larger
opportunities of observing the effects of sequestra which
have been left unextracted than perhaps any surgeon alive,
says, referring to his recorded cases : "I have given several
observations, *taken from among several hundreds*, in
order to show that the portions of non-extracted bone end
sooner or later by setting up eliminative action, which is
always painful, often dangerous, and at times fatal.   I have
also reported other cases in which immediate extraction has
been followed by positive cures, comparatively prompt.
These instances confirm the principles stated above.   Like
them, or even more, they confirm this truth, that the secondary sequestra, if they are not hurtful at the time when the
wound is received, or shortly afterward, become so almost
to a certainty at last.   They demonstrate the necessity of
removing them." Roux, Baudens, **Dupuytren, Guthrie,** and
nearly all the leading surgeons who **have seen many** gunshot wounds, repeat the same thing.   **I had** many times the
opportunity of seeing that these partially-detached fragments
seldom **lie** in the axis of **the** limb; so that if they did come

to enter into the new bone, they would be more a hinderance **than an assistance to its** assuming its functions, not only from their position, but also from their interposing between the principal sections of the fractured shaft, and preventing their contact and union. Their partially-necrosed condition makes them very liable to become separated by a future accident, and thus to be free to act more powerfully still as foreign bodies in the economy.*

Finally, considering the question in all its bearings, it must appear pretty evident that the removal of fragments must tend immensely to simplify the wounds under consideration, and therefore, that not only should all spiculæ which are entirely detached be removed as soon as possible, but that the same line of practice should be followed with regard to those which are so far detached as to retain but slight connections, and whose continued vitality must be doubtful; that this step should be acomplished by enlarging the exit wound; and that the practice is especially necessary in those cases where the femur is implicated, and a conical ball is the wounding cause.

The tertiary fragments, or those extensively adherent, should of course never be interfered with. Parts of these fragments may subsequently exfoliate, but at what period this may occur it is impossible to say. They may not appear for months, or it may be for years. Mr. Curling has lately made the observation, that necrosed portions of bone in compound fractures are longer of getting loose when they **are** connected with the lower, than when attached to the upper part of the shaft.

Any operative interference thought necessary for the removal of sequestra should be had recourse to at once, before inflammation has come on, or otherwise it will be

---

* Esmarch would seem to disapprove of the extraction of the secondary spiculæ, but the tenor of his further remarks tends to show the necessity for their removal.

more difficult for the surgeon, and not only more painful, but also more hurtful to the patient.

The few attempts that I saw in the East to resect parts of the continuity of the femur were certainly most unfortunate. Such a proceeding is manifestly much more severe and hazardous than that I have referred to above. The resections, however, did remarkably well in the leg and upper extremity.

In the classification of injuries which was followed in the Crimea, no distinction, unfortunately, was made between fractures in the upper, middle, and lower part of the femur, which prevents the discussion of several interesting points.

Although making every endeavor, I have only been able to find a record of three cases in which recovery followed a compound fracture in the upper third of the femur without amputation. In two of them the injury was occasioned by round balls, and the comminution was slight. In the third case I could not ascertain what species of ball had caused the injury. In one of these the patient, an officer of the 17th Regiment, was in the highest health at the time when he was wounded, (8th September,) and was of a peculiarly buoyant and hopeful temperament. The ball entered behind, and was removed in front, a little below the great trochanter, by Dr. Ward of his regiment. This patient received an amount of attention which it would have been quite impossible to bestow in the field under ordinary circumstances. He had a mattress constructed so that his wound could be dressed, and the bedpan introduced without disturbing his limb. He was wounded at a time when the comforts of camp-life were little behind those of home; and yet I have been informed that although his limb was in a very good condition when he left for England, the trouble it has since given him, and the deformed condition in which it remains, make it by no means an agreeable appendage. Another case was that of a soldier of the 62d, who was found a day or two after being wounded, lying in the dockyard stores of

Sebastopol, under the charge of the Russian surgeons. He
was discovered when the place was evacuated, and carried to
his regimental hospital, where he recovered. The fracture
in this case was in the lower part of the upper third. It
had been occasioned by a round ball, and the splintering
was not great. This man, however, was in the best health
when hit. He had just joined from England, and his injury
was comparatively slight. The third man may be said to
have had his limb consolidated, in so far as that a mass of
callus was thrown out, which cemented the bone; but he
died of purulent poisoning, and never left the Crimea. I
could not find out whether it was a round or a conical ball
which caused the fracture in this case. I know that the
French had hardly any recoveries. One was, however,
presented by the Baron Larrey to the Societié de Chirurgie
last May. This officer had been wounded in the upper
third, and the bone had consolidated.*   I never could hear
of any other except a Russian, whose greatly shortened and
deformed limb I often examined at Constantinople. This
man's thigh was quite firm, and had been allowed to unite
almost without treatment. There were probably a few
other cases, but they did not fall under my notice; although
during constant wanderings through the hospitals in front
and on the Bosphorus, I was unremitting in my inquiries
after such cases. I am certain, however, that although the
instances of recoveries were rare, they were yet not so exceptional as recoveries after amputation at the same part,
as will be afterward more particularly dwelt upon; and
thus it appears that, so far as the experience of this war is
concerned, we must conclude that, slight as the chance of
saving life is in any case, it is still our part to attempt consolidation in preference to amputation, when the fracture is
in the upper third of the bone. M. Simon, of Geissen,

---

* The records of the Val de Grace do not say what sort of ball
caused the fracture in this case.

draws a like conclusion from a review of all the reported cases of the injury; but he extends the doctrine to the middle third, in which I cannot agree with him, for reasons which I will afterward state. In the Sleswick-Holstein war, they preferred amputation to preservation in such cases. M. Hutin, in the Invalides, was able **to** discover twenty-four cases of recovery after compound fracture by gunshot **above** the middle of the thigh, but no case **of** recovery **after** amputation in the same part. This goes further to prove the position maintained above. In whatever way we decide, it is unfortunately too true that death will most commonly follow; but yet, when we do not operate, the patient may live in comparative comfort for several weeks, while, **in** the other case, he has to undergo a very fearful operation, and almost certainly dies within twenty **days.**

From the construction and limited range of the official returns, it is impossible to show in figures, what was, **however,** a well-recognized result of the **surgery** of the war, that though union did in rare cases follow compound fractures in **the** middle and **lower** third of the thigh, still **the** ultimate percentage of loss was greatly less when primary amputation had been performed than when limbs were saved, or tried to be preserved, or removed at a late period. When we take into consideration the fact so well brought out by the authors of the "Compendium de Chirurgie Pratique," and partly given in the note*—that we should, on the one side, calcu-

---

* "**If we** take 100 wounded," says the author of the article " Plaies d'armes a feu," "all of whom have received severe injuries of the extremity, necessitating amputation, and operate immediately —if, on the other hand, we take another hundred in the same condition, and wait to perform secondary amputation on those who survive the primary accidents, and then compare results—as far as it is possible to judge from observations borrowed from military surgeons, here is what follows in either case. Without hoping in the first case for a success equal to that of which S. Cooper, and Larrey, and other mil-

late those who die before the period for consecutive amputation comes round, as well as those who do not recover from it, and not merely those who die after being submitted to the operation—then the force of the teaching which inculcates primary amputation in these cases becomes much greater. Besides, as the cases which were retained for trial were always those in which the amount of injury was least severe, and the patients those most adapted for recovery, the presumption in favor of early amputation is the more decided. There can be little doubt that the chance of obtaining consolidation is greater in the lower than in the middle third, as is also the hope of recovery from amputation; so that, taking one thing with another, the experience of this war would lead to the conclusion, that when the thigh is fractured by a ball in the upper third, it should be saved, but that amputation should be immediately had recourse to in cases of a like injury occurring in the middle or lower thirds. Those fractures of a simple description, which at times present themselves, are not meant to be included in this remark, nor is it to be understood that, under more auspi-

---

itary surgeons speak, we may reasonably expect that the fatality will be here what it is in the greater part of other amputations; that is to say, that it will not pass the fourth or the third of those operated on. In the second case, on the contrary, if we admit the number of Bilsguers, who was so decided an opponent to amputation, we must expect to see the half of the wounded succumb to the primary accidents, such as gangrene, inflammation, abscess, etc.; 50 patients in the 100 will thus remain, retaining a wound which will call for secondary operation. When, then, the success surpasses all expectation, when we save 9 in 10 operated on, the number of those who survive will be yet less than if we followed the opposite practice; as, accepting the preceding hypothesis as exact, (and we think we have made them more favorable to secondary amputations than facts countenance,) there remains 70 to 75 surviving in 100 after immediate amputation, while there remains 45 or more after consecutive amputation. We also hasten to add that, in allowing for the moment the superiority of secondary over primary amputation, we have made an exaggerated concession," etc.

cious circumstances as to the condition of the patients and the means of treatment, better results than those we meet with may not follow the preserving of the limb. In fact, under ordinary circumstances, recourse should always be had to the steps I before spoke of, with regard to the removal of spiculæ in cases of fracture of the lower third, and then try to save the limb; but in a like injury of the middle third, the rule should be to amputate.

It is certainly very much opposed to the modern ideas of conservatism to condemn limbs without a trial, and I am fully aware how difficult it is to become persuaded of its necessity; but the unwilling conversion at last is made, though it is generally gained by the loss of several lives. The French surgeons in the East fully acknowledged the hopelessness of these cases; but the fatality of amputation **was,** with them, little behind that of preservation. This experience is as old as the history of war, and comes re-peated in renewed accents from every battle-field. Military surgeons are almost unanimous upon the **necessity of ampu-**tating in the cases specified, and most civilians who have had **an** opportunity **of** seeing much **of** these accidents **have** come to a like conclusion, as can be seen by the tenor of the communications to the Academy by the first surgeons of France. It would be mere **waste of time** to record **the** strong and decided verdicts **which have been given on** this point, and which find their summing up in the words of one **of** the greatest surgeons of any age or country, when Dupuy-tren says, in **one** of his clinical lessons: "I have repeated it often, and I repeat it for the last time, after the facts which I have observed, chiefly in 1814, 1815, and 1830, that **my** opinion upon this point **is unshaken.** In compound frac-tures from gunshot, in rejecting amputation *we lose more lives than we save limbs.*" The sagacious **Hennen** indorses **the** same view when he says: "I **am well** convinced the sum of human misery will be most materially lessened by permit-**ting** no ambiguous case to be subjected to the trial of pre-

serving the limb." Larrey, Guthrie, and in fact all the
leading military surgeons of modern times, proclaim the
same thing. That exceptions must sometimes be made, is
undoubted; but still they are only exceptions, and rare ones
too. Cases of compound fracture near the knee peculiarly
call for amputation, if the bone be split into the joint.*

The results which we obtained might most likely have
been more satisfactory if the army had made another campaign. Our bad hygienic condition deprived us of the improvements made in surgery during the last half century.

But, even in those exceptional cases which result in consolidation, the condition of the limb is not encouraging.
To this Guthrie bears strong testimony from his experience

---

* I cannot avoid giving the following remarks of M. Begin: "All
military surgeons have begun by wishing to preserve, but, as their
experience increased and their observation extended, they amputated
more, and they gain the conviction that they are right. At the outset of my career I amputated less than I did toward the end of my
service, as surgeon-in-chief of great establishments. There are certain cases, very often exaggerated, of wounded who pretend to have
preserved limbs which the surgeon wished to remove; I have been
present very often at the miserable death of persons who have refused
the operation, or who, they thought, would avoid it. The small number of the first, who boasted loudly, cannot compensate for those
much more numerous of the second, which caused me much sorrow.
And besides, how often are these preserved members not a pitiable
burden for those who carry them? Ask the surgeon of the Invalides
if he is not asked every year by some of these old soldiers to deliver
them from the parts which are an annoyance to them, and which
cause them inconvenience and incessant pain. I think it a great misfortune that our military surgeons should allow themselves to be
seduced by some of the assertions which you have heard; this for-
getfulness of the experience of their most illustrious predecessors
will cause certainly the loss of many men, which the art, exercised
with a more reasonable energy, might save." "I know that there
exist examples of recoveries with shortening, and fistulæ remaining
for years," says Baudens; "but to save two with fractured femurs,
and to heal them imperfectly, we will lose thirty, of whom fifteen or
more would have survived immediate amputation."

after Toulouse. M. Ribes, as is well known, failed to find
a single case of recovery, either after compound fracture or
amputation in the middle of the femur, among 4000 cases
which he examined in the Invalides at the period of his first
visit ; but during subsequent years he saw seven cases there
of "cured" compound fractures, five of whom died after
many years of great suffering arising from the injury, and
the other two be lost sight of, as they left the institution ;
but when last seen they were in a grievous plight, and he
says, "it is probable that these two soldiers died from the
effects of their accidents, and if they did not, their condition
must be greatly still more wretched." In all the seven cases
there was union certainly, but it was attended by much deformity, necrosis, and caries.   Long years of suffering, constant abscesses, exfoliations, atrophy, sensitiveness to the
slightest atmospheric change, shortening and deformity, the
development of phthisis, if it be in the constitution,—these
are among the results of a "cure" of a compound fracture
by gunshot in the middle of the thigh.*

---

* In the **Punjab, and** other Indian campaigns, I have been able
to find **the details of 24** cases of compound fracture of the thigh,
(parts not specified,) in which the attempt at saving the limb was
made. Of these, 14 died very soon ; but of the ultimate state of the
remaining 10, or whether they continue to survive, **I** find no notice.
Dupuytren, in 1830, lost 7 out of 13 cases treated by him.   Mal-
gaigne, in 1848, lost 3 out of 5, all being select cases, and those not
adapted for immediate amputation.   Baudens, in one series of 60,
which he mentions in his book, amputated 15 immediately, of whom
13 survived ; 20 were amputated late, of whom only 4 recovered.
The remaining 25, although tried "avec obstination" to be saved, all
died miserably except 2, who retained **"a** deformed member, **unfit to**
fulfill its functions," and which, he says, they would **willingly part**
with.   "Taking **a** retrospective **view,"** says Bell, **"we see in true**
perspective all the dangers **of a nine-months' cure, which is but a**
weary travel, step by step, betwixt **life and death.   In this** view we
see the dangers of frequent fevers, wasting diarrhœas, foul and

Finally, then, let me repeat the conclusion; that under circumstances of war similar to those which occurred in the East, we ought to try to save compound comminuted fractures of the thigh when situated in the upper third; but that immediate amputation should be had recourse to in the case of a like accident occurring in the middle or lower third.

Many of the fractures of the leg were so severe as to call for early amputation. Severe shell or round shot wounds seldom leave much hope of saving the limb; but in a large number, however, of very unpromising cases, the leg was preserved. A great deal was done in the leg in the way of **removing** fragments. Guthrie says they can be extracted "**to** almost any extent and number," and he directs us, if necessary, to saw off irritating parts of the ends of the shafts. If one bone only be broken, and the loss of substance in it is not great, the case will be the more promising, as the unbroken bone keeps the fractured one steady and the soft parts in place. It is when a scale of the bone, however thin, remains, as we occasionally see it in shell wounds, that the best results in the way of cure are obtained. Such was the case in a most successful instance of repair, in a man of the 20th Regiment, under the care of my friend Dr. Howard of that regiment. I **relate** it, because it may be looked on as an example of a class of cases which were not uncommon. **A piece** of shell **struck** the edge of the left tibia, and de**stroyed the greater part of the** thickness of its shaft, from

---

gleety **sores: some dying suddenly of gangrene, some wasted by the** profuse **discharge and successive suppurations, new incisions, and** unexpected discharges **of spoiled bones: we see those who recover** halting **on** limbs so deformed **and cumbersome that they are** rather **a** burden than **a help. In the very moment that we hear of** such a **cure, we** know how much the patient **must have** suffered, and how **poorly he has been cured;** and we can, from **the** long sufferings of **those who** escape, tell but too truly how many must die."

below the tubercle downward for about three and a half inches. The fragments were removed at the time of the accident, or afterward, as they became loose; the posterior shell of the bone being, however, entire, was carefully preserved. Four months afterward this patient was sent to England with a strong and useful leg, whose only change was a slight bending outward—a condition which generally remains in these instances. This case was just such a one as presents the best hope for a good result. I by no means would infer that some most excellent recoveries did not take place when resections were performed of pieces, including whole thickness of the shaft of the tibia; but they were the much more rare, and infinitely more tedious than cases like the foregoing. When the leg is fractured low down near the ankle by a ball, the accident is much more grave than when it takes place at the middle of the limb. I have exceedingly seldom seen a case recover in which the tibia was split into the joint.

The free anastomosis which exists between the vessels of the upper extremity, the large supply of blood which they convey, the ready development of a compensating circulation, the less drain there is on the system during the period of suppuration, and the less call there is for the patient to retain a constrained and irksome position during cure, render many things practical in compound fractures of the upper extremity which could never be attempted in like injuries of the lower limb. The injury, indeed, would need to be very extensive before we would think of performing amputation at an early period in gunshot wounds of the arm; as, unless the vessels are destroyed, there are many **most** dreadful and hopeless-looking accidents from which **the arm** will recover; **and,** besides, **secondary** amputations **are so** successful, and resections so **often sufficient to fulfill** the necessary indications, that primary amputation is never performed in the upper extremity except under the most des-

perate circumstances.* Stromeyer recommends the trunk
to be made the splint in treating these cases, so as to do
away with all that fear of motion in the fragments which
exists if they are treated in the usual way.   Unfortunately,
however, as pus commonly burrows, and has to be evacuated on the inner aspect of the arm, it is difficult to carry
such an idea into practice.   Pirogoff, it appears, was so
displeased with the results of his attempts to cure fractures
of the upper extremity, in the Caucasus, that he was disposed to submit them all to amputation.   The world will
learn with interest whether his experience in Sebastopol has
not been more favorable.

The results, with regard to fractures of the **forearm**, do
not tell the whole truth, as there is no provision made in the
returns for showing double injuries; many cases are made
to appear as having ended fatally, from these and other
comparatively trivial injuries, which **were**, in truth, the result of a complication of accidents, of which this was the
one chosen for registration. I have known this occur often.
Fractures of the forearm, when not combined with other injuries, turned out most satisfactorily.   Hardly a case came
under my notice which did not do well, even although the
comminution of the bones was very considerable.

As to the treatment of **compound fractures** little remains
to be said beyond what has been already hinted at, or what
is commonly pursued.   Perfect fixture—a fixture so well secured as, if possible, never to be disturbed during the pro-

---

* The **following is a curious** instance of recovery from a most
hopeless-looking injury.   It is related in one of the Indian regimental reports in the War-office. A soldier received, in the Khyber
pass, a sword-cut which divided his arm, bone **and all**, with the
exception of the vessels and **nerves**, and the **muscles** on the inner
side.   He also received another **wound, which laid bare** the spine
and ribs; yet he recovered, the bone of the arm uniting.   He died
afterward of another accident. Two somewhat similar instances, one
from Percy, are related by Ballingall, pp. 343-4.

cess of consolidation ; plenty of fresh air ; the free discharge
of pus obtained by judicious and early incisions and by position, and *not* by manipulations of the injured part ; and the
administration of tonics and nourishment, but as little strong
stimulation with brandy and wine as possible,—these comprise all the chief points in the treatment.

Purulent absorption has been the cause of death in the
vast majority of those compound fractures which ended
fatally. Pus, occupying both the chief veins and the inter-
stices of the bone, was commonly found, and purulent deposits in the lungs very generally existed. I do not think,
looking at the question as a whole, that our experience
would lead us to subscribe to Velpeau's doctrine, that " purulent absorption is more common among those who undergo
amputation than among those who have severe suppurations
and preserve their limbs." Hectic, the renewal of old enteric disease, and cholera carried off many of our patients
under treatment for compound fracture.

The results which followed the treatment **of gunshot**
wounds of the hand and **foot were very satisfactory in most**
instances. Balls perforating either created a great deal of
destruction, but the rapair was not slow. " The talent of
preserving" was well shown in the Crimean hospitals in
these instances, and in general the results rewarded the **en-**
deavors made **to** save the **member.**

It is remarkable how few sequestra separate in gunshot
wounds of the hand, even when the shattering of the bones
has been great. The extrusion of any large piece of bone
seldom occurred, so far as I saw. In gunshot wounds perforating the foot, the most marked feature was the **great**
swelling which followed, and the **extreme pain which this**
distention generally **caused.** How far **the rapid cures** obtained in the field may **remain permanent, I am at a** loss to
know ; but I fear that not **a few of the cases** " patched up"
and sent home may have **to** undergo operation at a subse-
quent date.

In dealing with gunshot injuries so severe as to demand
operation in the field, we can often save more of the part of
the hand or foot than usually after accidents in civil life.
The soft parts are seldom so much destroyed, in proportion
to the injury inflicted on the hard tissues, by a musket-ball as
by a wheel of machinery; and thus we are not called upon
to remove so much of the member in order to secure a good
covering for the hard tissues.

CHAPTER XI

Gunshot wounds of joints form a group of cases **most** interesting to the surgeon. " As for a wounded joint," says John Bell in his treatise on gunshot wounds, " we may take the united experience of all surgeons, which has established **this as the true** prognostic, that *wounds of the joints are mortal.*" Without, however, being so sweeping in the condemnation of such cases, it must be affirmed that no class **of** gunshot injuries prove more uncertain in their results, or are more commonly followed by disastrous consequences.

The gravity of gunshot wounds of the joints will depend chiefly on the size and construction of the articulation, the extent of the injury, and the attention received by the patient shortly after being wounded—especially the means of treatment being at hand, and not necessitating long transport.* As a very grave amount of destruction may be in-

---

* Mr. Alcock classes cases of wounded joints, with reference to their results, under three heads: those treated under 1. "*Favorable circumstances.*—Cases admitted into a large, well-organized, and commodious hospital an hour **or two** after **the** injury was inflicted, **and** there treated **to** the end, under the same medical superintendence, and with all essential means for good treatment. **2.** *Partially unfavorable circumstances.*—Cases not immediately received **into a well-organized hospital, subjected to some leagues** of **transport, or passing** part of **the first period in a field hospital with deficient** means, or received into **a** permanent **hospital with lax discipline. 3.** Cases treated under *unfavorable circumstances,* **or those** admitted into crowded hospitals with epidemics prevailing, means either personal or material not fully adequate; with cases of wounds inflicted after

flicted on the articulating extremities of the bones without much external appearance of such mischief, we are often deceived in our early examination of these cases; and this is one reason for delaying the adoption of decided measures though delay so frequently proves fatal.

The wound of a **ginglymoid articulation** is, as a general rule, more severe than that of a **ball-and-socket joint**, chiefly from its more complex structure. Larrey noticed how often **tetanus** was caused by wounds of these joints, and **every** surgeon can testify to the extremely severe symptoms which follow their injury.

Although it is true in general that a mere fissure extending into a joint may not be followed by serious results, still it is no less certain that even such apparently trivial accidents are often followed by the most disastrous consequences.

It is a matter of much moment to possess a decided opinion upon the treatment of gunshot wounds of the joints, as in no class of cases is prompt action so much called for, and none in which, by the parade of a few successful cases, is the mind of the surgeon more apt to be misled. If, on seeing a case, we were able to decide what remedies were demanded for its management, then possibly much suffering and no few lives would be saved.

Gunshot wounds of the **neighborhood of joints** require much attention, not only from the fear of secondary implication of the articulation, but on account of the stiffness **which is** apt to ensue in it from long disuse during the **period of cure.** Artificial motion should be begun early in these **cases.**

---

a reverse in the field, or long subjected to the deficient means, discomforts, and imperfect discipline of temporary or field hospitals, with one or two days' subsequent transport to the permanent hospital stations." He adds: "The evidence of these statistical results (those having reference to the above points) is too striking to leave any doubt whatever as to the influence which these circumstances exercise, totally independent of the constitution of the patient and the degree of injury."

The hip is too deeply placed, and too much protected by the surrounding parts and its own form, to be often penetrated by a ball; but when it is implicated, the destruction is commonly so great as to render operative interference in some form imperative. Alcock lost three out of four cases in which this accident occurred, and in the fourth case, "where recovery took place, the joint itself, there is some reason to suspect, was but remotely affected." Occasionally a round ball becomes impacted in the head of the femur, and may cause only a partial fracture of its neck. It is not easy in either of these accidents, however, to recognize the injury at first, as no sign of displacement or crepitation may be perceived. This is, however, rare; but the following is one case of this description. It is related in the register of the Depot Hospital, at Colaba, in the archives of the medical department: Alexander M'Phail, aged 33, wounded at Dubba, 24th March, 1843, *by a match-lock ball*, which entered a little above the great trochanter of the right limb anteriorly, and was lost. His leg became powerless. On coming to Colaba on the 26th April, he did not complain of much pain, except when the joint was moved. Slight fullness over the hip was the only symptom of injury. Leeches and counter-irritation were employed, and he seemed to get better. On May 6th he was attacked with **trismus, and died on the 9th.** The ball was found imbedded in the head of the femur, which, with half of the brim of the acetabulum, was shattered, and the capsular ligament formed the sac of an abscess which contained a considerable quantity of pus and spiculæ of bone. The orifice of the wound, it is added, had closed some time previous to death.* Larrey mentions the case of an officer

---

* Preparation 2604, in the **museum at Fort** Pitt, **was, I** believe, obtained from this patient, as the description of it in the catalogue is as follows: "A *match-lock* ball firmly lodged in the head of the femur. It entered opposite to the **trochanter** major, and passed

24

wounded in Egypt, who received a ball in the neck of the
femur. The wound closed, and, twenty years afterward, on
the death of the patient from disease of the chest, the ball
was found impacted in the bone.

The **knee** when penetrated by gunshot presents an injury
of the gravest description. Taking much interest in cases
of this description, I visited every one I could hear of in
camp, and can aver that I have never met with one instance
of recovery in which the joint was distinctly opened, and the
bones forming it much injured by a ball, unless the limb
was removed; yet the returns show several recoveries after
such wounds, some of which, at any rate, I cannot but think
are founded on error. I have conversed with many surgeons
of large experience on the subject, but never heard of any
case recovering without amputation, in which the diagnosis
of fracture of the epiphysis was beyond doubt; yet such
cases have been put on **record.** I remember one case,
probably included among the recoveries, in which a ball
passed near the joint, causing some effusion and swelling in
it, with no constitutional disturbance whatever, and resulting
in the man's return to duty within a fortnight, but which
the surgeon in charge put down as a penetrating wound,
remarking (as he well might) on the curious immunity from
constitutional or severe local symptoms which had marked
the case.

The following is a very interesting case, and certainly
one of the most difficult to explain of any with which I am
acquainted. I never saw the patient, but the details have
been **kindly sent me** by Deputy-Inspector Taylor, from
Chatham: "Private George Hayes, aged 31, 47th Regiment, was wounded at the Alma by a grape-shot, which
entered on the outer side of the ligamentum patellæ, and

---

through the brim of the acetabulum. The wound in the skin soon
cicatrized, but the patient died of tetanus six weeks after the receipt
of the injury."

passed upward through the knee-joint, shattering the patella in its course, and making its exit at the anterior aspect of the thigh about its middle, partially fracturing it. The greater portion of the patella was removed in the course of treatment, as well as various fragments of the femur (exfoliations?); but firm union of the latter, as well as anchylosis of the joint, fortunately took place. At the time of his discharge he could sustain his weight upon the limb, and could walk about without crutches." I saw another case very similar to this at Scutari, in 1855. In this instance, the ball had struck the man when he was about to kneel, and apparently fractured the head of the tibia. The ball **was** removed from the anterior part of the thigh. Scarcely any bad symptom followed, except that the joint swelled, was painful to the touch, and ended by losing part of its motion. If the articulation escaped the passage of the ball, the case was very curious.

The **round ball** sometimes penetrates the **lower end of the femur** or the **head of the tibia** without causing splintering, or opening the joint, or at least with an amount of injury to the capsule which is very slight; and such cases may recover, and so shake our conclusions about others of a less anomalous character. Balls, too, may pass very close to the capsule and yet do it no harm, though these cases are put down as penetrating or perforating wounds of the joint.*

It is undoubtedly often very difficult to know whether the joint has been opened or not, particularly if the ball is a

---

* Alcock thinks that if a ball do not absolutely project within the articulation, or if the foreign body be smooth, and not project much beyond the articulating surface, the limb may **be saved.** Independently of the fact that the cases **are infinitely** rare **in** which a ball—especially a conical one—can thus **penetrate, without** causing grave and irremediable fracture of the bones entering into the articulation, it is not at all consistent with my observation that such cases can be saved. One case mentioned in the text illustrates this.

small one, as was the case in one instance afterward mentioned; and it very often occurs that the missile has run superficially under the integuments, or coursed round the bones, when it appears to have passed through the articulation. It is to be remembered, also, that the swelling of the joint may be merely the result of a bruise, or of the extension from the neighborhood of the inflammation which has been caused there by injury, and is thus no sign of direct wound of the joint.

Another point which renders these injuries difficult of recognition when the bones are not much implicated, is the length of time which may intervene before the appearance of severe symptoms. A week may pass, and yet both the local and constitutional symptoms may be very slight. Sooner or later, however, the well-known signs of joint-injury are set up, sometimes with great rapidity and **severity.**

It is not difficult to understand the peculiar progress and fatal results of gunshot wounds of the knee, when we consider how sensitive to injury are shut cavities when inclosed by such a delicate membrane as the synovial lining of the knee, and how feelingly such cavities resent the introduction of air within them; how rapidly they degenerate under the effects of this air; what a mass of closely-compacted tissues become implicated when disease is set up in such an articulation; **how** it is that bone, ligament, and soft parts participate in the injury; how wide the bony expanse is which enters into the formation of the joint; and what a large surface is presented for purulent absorption and transmitting inflammation, as well as how difficult it is for foreign bodies or morbid secretions to obtain free exit. These are the chief causes why the injuries under consideration are so often followed by dangerous and fatal results. In civil life, wounds opening **the** joint are commonly caused by cutting instruments. Foreign bodies are seldom introduced, and the bones entering into the articulation are little if at all injured.

The wound, being carefully closed, often adheres, and by appropriate treatment little mischief may follow. But if a ball be the wounding agent, foreign bodies are almost sure to be introduced from without, or created within by the splinters. The ball's track must suppurate before it closes, and it cannot be shut up and retained without the hazard of pus accumulating in the cavity; air thus gets admission, and works destruction. Foreign bodies cannot be extracted by so small an opening from a cavity of such a construction: and thus these gunshot wounds of the joint, though often apparently very trivial injuries, become the most serious almost of any which can be presented to us.

The primary dangers of these wounds are not great. It is in those which are set up afterward that the chief hazard exists. The long and wasting suppuration, the tedious and dangerous abscesses, and the purulent poisoning are the principal sources of alarm. These abscesses are most curious occurrences in knee cases. They appear almost invariably among the muscles of the thigh; and while they may remain long unnoticed, they give rise to the utmost trouble and danger. They burrow along the bone, often stripping it of its covering, and yet are seldom apparently in connection with the joint. The escape of some small amount of the acrid secretions into the superficial or deep cellular membrane sets up renewed inflammation and suppuration there, and thus abscesses form whose connection with the original depot it is difficult to trace. These collections almost always occur in the thigh in-preference to the leg. At a late period of the case, the joint puts on all the appearances of white swelling—an observation first made by Dr. John Thomson.

Military surgeons of all times have recognized the **neces**sity of removing the limb early **in these cases** when the articulating ends of **the bones have been fractured by a** ball, and the experience of the late war fully **bears** out the practice. French and English surgeons were, I think, agreed on this in the Crimea. In December, 1854, I saw upwards of forty cases in the French hospitals, and all died except those primarly amputated. I have heard incidentally of one case occurring in their army which recovered, but have failed to learn its details.* It is certainly very disheartening, as well as humiliating to professional pride, to think that we cannot save such cases without amputation. The very small amount of visible destruction which is so often present; the slight complaint of pain or appearance of disturbance which frequently exists at the period when the limb ought to be removed in order to insure success; the very pardonable unwillingness of the patient, especially if he be an officer, to submit to so dreadful an alternative, where there is, to him, so little apparent danger,—all render difficult the adoption of those measures which a dire experience has shown to be necessary, for that amputation is our only resource all are agreed.

Guthrie has seen no case recover in which the limb was not removed. Larrey reports some, but they were instances of slight injury. Esmarch, from the fields of Sleswick-Holstein, says : "All gunshot injuries of the knee-joint, in which the epiphysis of the femur or tibia has been affected,

---

* In the Indian Reports I have been able to find the particulars of nine cases in which the knee was penetrated, but the injury was apparently so slight as to lead the attendants to try to save the limb. Every one died. Alcock has stated the proportion of cases in which the articulations are wounded, to other gunshot wounds, as between 4 and 5 per cent., nearly one-half of which were of the knee. Of 65 cases in which an articulation was primarily affected, 33 recovered, **21 with loss** of a limb, 32 died, 18 without amputation. "It is quite evident," he adds, "that if the 18 cases of death without amputation, and the 14 cases of subsequent amputations, (assuming them to be unfavorable causes for treatment in the first instance,) instead of being treated, had immediately been amputated, we should then have had for result, not a loss of 25, but of one-third, which is the loss from primary amputation. Two-thirds, therefore, or 16 out of the 25, would have been saved." Of 35 cases in which the knee was more or less implicated, 22 lost their lives, and of the remainder, 8 lost their legs. "After such results, it is little to say that the 5 who recovered preserved good and useful limbs."

demand immediate amputation of the thigh. It is a rule of deplorable necessity already given by the best authorities, and which our experience fully confirms."

I have often contemplated the laying of the articulation freely open at an early period in these cases, so as to permit of the extraction of all foreign bodies, and the free escape of the pus which must afterward be formed, the retention of which is undoubtedly one great source of danger. This might be attempted even although it were necessary to lay the whole front of the joint open by an incision similar to that for excision.* The joint has been frequently widely laid open by cutting instruments, both primarily and for disease, and most satisfactory cures have been obtained.†

If, however, the attempt is to be made to save the limb, **the** most rigid antiphlogistic treatment must be followed. Local bleeding by leeches, and the application of cold; the avoidance of all local remedies which are of a relaxing nature; the perfect fixture **of** the articulation, and the **ab**sence of all pressure; as well as the early evacuation, **by free** incision, of abscesses, and of matter if it form within the joint,—these **are the leading and** evident indications to be followed. Hectic, with its common accompaniment, diarrhœa, purulent absorption, with secondary implication of internal organs, and tetanus, are the causes which most commonly bring about a **fatal result.**

The presence of the articular cartilages would be of little moment, as they soon disappear; and if the bones were kept **in close** contact, firmly fixed, and all discharges allowed freely to **escape,** there is no reason why most favorable results might not be obtained. Such **a** step is in **no way so**

---

* At the time the above was first **written, I had** not seen Stromeyer's book, **and** did **not know that the same idea** had occurred to **him,** or that, in the **only case in which he had** practiced it, the results had been most encouraging.

† See especially a paper, **by** Mr. Gay, **in the** *Lancet* **for** October 25, 1856.

severe as excision of the joint, and yet how successful has
this been! The great sources of irritation and danger would
be done away with, and if we had a healthy patient to deal
with, I 'cannot see why we should fail.

If amputation is thought of, the sooner it is undertaken
the better, as when the operation is performed late, after
inflammation and suppuration have been for some time
present, the results are very unfavorable. The joint, when
opened, presents most of the characteristics of chronic disease, plus the immediate injury to the bone — cartilages
eroded, synovial membranes degenerated, and the products
of inflammation effused into the cavity.

In those cases which are occasionally saved, the cure is
very slow and very unsatisfactory.* They occur only in

---

* Alcock says, with regard to the boasted *cures* after injured joints:
"By a limb saved I do not mean one with the wounds healed, having
nevertheless the extremity contracted, bent, motionless, or otherwise
useless. Cases, by a loose kind of phraseology, are often termed
'limbs saved.' The object of saving a limb is that it may be useful.
If this is not the result, the member by merely hanging to the body
of the patient is lost in my estimation as truly as if amputated, but
with the additional circumstances of being converted into a source of
misery to the sufferer, an impediment to the free motion of the rest
of the body, and often a cause of irremediable bad health. Such
cases I hold to be among the worst specimens of bad and injudicious
surgery." It is much to be regretted that surgeons cling so much to
a few cases of recovery, and shut their eyes to the vast mass of instances in which the attempt to save life and limb has failed, forget-
ting that "Les miracles ne peuvent servir de base à aucun jugement."
John Bell puts this with great force: "Thinking only of this wonderful recovery, the surgeon willingly forsakes an uncomfortable
rule to lay hold on this one glimpse of hope, while, indeed, if he
reasoned fairly, he would perceive that the exception should be lost
in the fullness of the general rule, and not the general rule disturbed
by the exception." Alcock, too, says: "In the class of injuries
under consideration, this danger is most especially evident. Many
are the extraordinary and most unlooked-for successes attending the
treatment of forlorn cases of injured joints. Were general rules or
principles of treatment to be founded on these cases, which are but

those instances in which the bones have been slightly injured
and the patients possessed of a first-rate unimpaired constitution, and when the means of treatment are of the most
perfect description.

The particulars of cases in which the knee-joint has been
penetrated by balls are so similar, that I will only give a
few in detail. The first case was under my immediate
notice, and from its presenting what might have been considered the most favorable features for conservatism, it gave
me much interest. An officer of the 63d Regiment, aged
19 years, was accidently wounded by his own revolver during
the reconnoisance in force at Kinburn, on the 21st October.
He was half reclining on the ground at the time of the accident, his left leg stretched, and his right knee half bent.
His pistol was in his right hand, and close to his limb. The
muzzle was directed downward, and obliquely inward, toward
the middle line of the body. The ball, a small conical one,
weighing four drachms, entered at the outer and superior
surface of the lower third of the right thigh, about three
inches above the border of the patella, and lodged. The
wound appeared to lead into the cavity of the joint, and an
indistinct grating was said to have been communicated to
the hand on moving the patella from side to side. Some
bloody greasy fluid escaped from the wound. On the
patient's coming into camp, a few hours afterward, I first
saw him. There was then considerable swelling around the
wound, but the motions of the joint were free, and unattended

units among thousands giving contrary results, and were no reference made to these greater numbers which enlarged experience
shows must perish in vain attempts to save limbs, an immense sacrifice of life and increase of human suffering would inevitably follow."
Guthrie, after emphatically protesting against our being guided by
these exceptional cases, says: "If one case of recovery should take
place in fifty, is it any sort of equivalent for the sacrifice of the other
forty nine? Or is the preserving of a limb of this kind an equivalent for the loss of one man?"

by almost any pain, and there was no swelling whatever of
the articulation. There was one small spot over the head
of the fibula, of which he greatly complained. Two tracks
seemed to lead from the wound. One ran toward the inner
side of the joint, and the other went along its external
aspect; both were quite superficial. The position of the
ball could by no means be determined. The patient's youth
and strength, the absence of positive proof that the articulation had been opened, together with the possibility there
existed of the ball having been deflected, and having passed
down by the track along the external aspect of the limb
and lodged about the place so loudly complained of, in the
neighborhood of the head of the fibula, made us determine
to wait. The most decided measures were immediately
taken to ward off inflammation. The joint was fixed, and
as he was taken on board ship and put under the immediate
charge of my friend Mr. White of the 3d Dragoon Guards,
every care was bestowed on him. On the 24th there was
some swelling of the joint, accompanied with pain. The
wound of entrance was beginning to suppurate well. Synovia had not been seen to escape. His pulse was 78 and
soft, and his secretions were natural. The penetration of
the joint was believed by all, still an attempt to save the
limb was determined on. The usual local and constitutional antiphlogistic remedies, including evaporating lotions,
fomentations, leeches and cupping, antimony and calomel,
were diligently put into requisition. By the 30th, the joint
was much swollen, and very painful on pressure. A spot,
about the size of a shilling, over the head of the tibia, was
exquisitely tender. He was feverish at night, when the pain
was always much exacerbated. An abscess formed among
the muscles of the thigh, and continued to suppurate profusely after being opened. It had no apparent connection
with the joint. Slight hectic set in. The pain in the joint
became lancinating and throbbing to a most harassing degree, particularly over the head of the leg bones and over

the patella. The articulation assumed all the appearance and feel of a joint affected with white swelling. This was the state of things by November 19th, when amputation was finally decided on, and performed in the middle of the thigh, the state of the limb not allowing of its performance lower down. On the 1st December he was transferred to the Castle Hospital at Balaklava, at which date the stump was suppurating kindly, but an erythematous blush overspread the integuments of the limb as high as the hip. This disappeared in a day or two, and the stump cicatrized, all except a small part in the center. His strength seemed to improve. On the 4th, he had a slight rigor, which was repeated twice daily. He gradually sank, had cold sweats, dyspnœa, diarrhœa, and died on the 8th. The end of the stump was hollow, and contained much pus, and half an inch of the end of the femur was dead. The lungs were congested, but beyond this no particular appearance was observed. The vessels of the stump seemed healthy. In the removed limb, the tissues around the knee-joint were found much engorged, the articular cartilages and ligaments were quite disorganized, and the cavity filled with turbid purulent fluid. The ball lay below the patella, in the intercondyloid notch. Its pressure on the end of the femur, lower surface of patella, and head of the tibia was marked by complete erosion of the cartilages on these points. The bones were not otherwise injured. This case is highly interesting, not only on account of the difficulty found in detecting the ball, lying in the position it occupied, but also from the uncertainty which marked the line of procedure at the outset, the long absence of serious symptoms, the smallness of the ball, and the fatal result, notwithstanding the slight injury to the bones. If a free opening had been made early into the articulation, might we not have saved life and limb?

Miller, private, 31st Regiment, was admitted into the general hospital on the 9th July. While his left leg was coming forward, as he was marching down to the trenches,

he was struck by a piece of shell over the lower end of the
femur and the external surface of the knee-joint. The
wound was about four inches long, little lacerated, but deep,
and opened the joint. The wound was carefully closed by
suture, the limb fixed, and cold was applied. Inflammation
was violent by the 16th, notwithstanding the employment of
every means to moderate it. The wound opened and synovia escaped freely. The constitution did not apparently
sympathize much for many days after the local inflammation
was considerable. Pus was poured out freely by the wound,
symptoms of pyæmia rapidly set in, and he ultimately died
on the 3d of August. The external condyle was splintered
into the joint, the cartilages were eroded, but little fluid existed in the articulation. If the joint had been freely laid
open after the wound had failed to adhere, might not a
better result have been reasonably looked for?

Shell wounds of the knee are as a whole not so dangerous
as bullet wounds. They frequently merely cut the soft
parts open; or if they injure the bone, the larger aperture
which they leave acts beneficially in permitting the free discharge of secretions. I have known many shell wounds in
the neighborhood of the joint, and not a few in which the
articulation was opened and even the bones injured, ultimately do well, so far as saving the limb goes; more or less
anchylosis following. The following short notes of three
cases of this description were kindly sent me from Chatham,
and are of more use as bearing on this point than other
cases the outlines of which I possess, as they record the
state of the patient some time after being injured : Patrick
Madden, private, 49th Regiment, was struck by a piece of
shell on the left knee, when in the trenches, on the 30th of
May, 1855. The joint, if not opened, was very gravely injured. He was two months under treatment in the Crimea,
with his articulation much swollen and inflamed, and when
he reached England the wound had healed. He was invalided on account of "swelling, pain, and weakness of left

knee," the joint being partially anchylosed and its tissues thickened.

Private John Dwyer, 49th Regiment, aged 29, was invalided on the 23d of April, 1856, for "partial stiffness of the left knee," occasioned by a shell wound which partially fractured the patella.

Private James Callaghan, 95th Regiment, aged 21, was invalided at the same station, on the 14th of April, 1856, for "anchylosis of the left knee." This man had received a shell wound of the joint, but no mention is made of any fracture of the bones. The articulation was still painful at the period he was invalided.

A paymaster-sergeant, belonging to the 38th Regiment, a dissipated, nervous man, was admitted into the general hospital on the 18th of June. While kneeling, he was struck on the knee of the limb not on the ground, by a piece of shell, which was supposed to have lodged. The wound appeared to lead into the cavity of the joint, and much injury of the head of the tibia had been evidently produced. The patient would not consent to the removal of his limb, and being a non-commissioned officer, his desire was complied with. The limb was slightly bent, and laid on a pillow, while local and constitutional remedies were promptly applied. For eight or ten days no disturbance, local or constitutional, supervened. The joint then began to swell, became glazed and painful, and his stomach became irritable. The pain was chiefly confined to a point on the inner side of the joint. He went on, one day better and another worse, the joint always becoming more hopelessly diseased, till July 15th, when the limb was removed. For a time he did well, but ultimately sank in the beginning of August. The head of the tibia was much injured and split into the joint, while part of its shaft was driven upward into the cavity and indented the condyle of the femur. A piece of the head of the tibia was also driven downward and was impacted in its own shaft. The articulation was filled with a dirty pink-

25

colored matter, but the cartilages were not diseased. The
flaps had adhered to a considerable extent, but within them
a large cavity was found filled with pus, decomposed tissue,
and blood. This cavity extended up along the external surface of the bone to the trochanter major. This case was an
example of a large class in which amputation was performed
late for gunshot wounds of the knee, and in which this large
depot of pus had formed. These are cases in which adhesion of the flaps by the first intention should never be attempted, but the utmost facility given for the escape of
matter.

I have seen only one case in which the **patella** being
fractured by a ball, the joint was not at the same time
opened. The bone was in that case "starred," but the ball
did not lodge. The subsequent inflammation of the joint
was slight, and the recovery good, the motion of the joint
being, however, considerably interfered with.

Penetrating wounds of the ankle generally did well,
although they required long treatment. This is opposed to
the usual experience of such injuries, but very much seemed
to depend on attention to two things—first, that the articulation was rendered perfectly immovable ; and, secondly,
that one or other of the wounds was so enlarged as to allow
of the free escape of all discharges. If the original wounds
were large they generally did best, as surgeons are unfortunately averse to render them free if they are not so originally.* The truth of this remark I have had ample oppor-

---

* This observation is as old as Ledran, who says: "Sur ce prin-
cipe j'ajouterai qu'une playe dans laquelle toute la moitié d'une
jointure serait emportée doit être regardée comme beaucoup moins
dangereuse qu'une playe qui la perceroit de part en part." "Contrary to the general impression," says Alcock, "I am strongly
inclined to the conclusion that injuries to joints are not fatal in proportion to the extent of surface laid open. The most dangerous of
these wounds I believe to be punctured, or such wounds as a musket-
ball creates—a small lacerated and contused opening, with more or
less mischief to the internal parts."

tunity of verifying. Stromeyer is of opinion that if there is much destruction of the external malleolus, we should remove the limb, as the foot takes on the appearance of valgus, **and** is useless. I have not observed this result.

The **shoulder-joint** has recovered well in several cases which I have noticed, where a ball has passed through part of it, and even in cases in which a good deal of the head of the bone has been destroyed. I suspect, however, that the after-consequences are not always so encouraging as the rapid healing would lead us to expect. I have under my charge, at the present moment, an officer who was wounded at the cavalry charge at Balaklava, by a rifle-ball which shattered part of the scapula and the head of the humerus. Nothing was done in the way of extracting the broken fragments at the time of injury; and accordingly, besides the hazards of a very long and tedious suppuration, during which many considerable spiculæ have been extruded, and the formation and evacuation of long "fusees" of pus, he is yet subject to a constant recurrence of these purulent formations, and to the exfoliations of pieces of bone which seem to be set loose by the least overexertion. His joint is quite anchylosed. If excision had been practiced early, might not a more useful limb have been retained, and much annoyance avoided ?

The much greater simplicity and superficial position of the shoulder than of the other articulations, cause it both to suffer less and to be more manageable when injured. Balls sometimes pass very close to the capsule without opening it, or, at any rate, injure it but slightly. Of this, I believe, I have seen several instances. Larrey has recorded a case in which a round shot passed across the shoulder-joint ; and although it only abraded the skin, it yet shattered the head of the humerus, the scapular end of the clavicle, as well as the acromion and coracoid processes. This man was saved by the excision of the destroyed bone. He tells us he saved many cases in which the opening into the joint was not **great**.

If a ball remains impacted in the **head of the bone**, as it sometimes is known to do, then the sooner it is got rid of the better, as caries of the bone, disease of the joint, and either amputation or death will follow. One case occurred at Scutari, in which the ball was found after death firmly impacted in the round head of the bone. I find the report of a similar case as having occurred in one of the regiments serving in China in 1841–42. In this case also the ball was not removed, necrosis was caused, and the patient died of exhaustion on the fiftieth day. Malgaigne, however, reports a case in which a ball had been so englobed, and no disease whatever caused in the bone, where a considerable cavity contained the ball. Abscesses and fistulous tracks are the things most to be dreaded in all cases in which the shoulder-joint is implicated in gunshot wounds.

In the following case of **penetration of the elbow**, the distinction between a wound caused by a **sword-cut** from one made by a **ball** was well shown. If a ball had passed across the articulation, fracturing the bones, excision would have been called for: A dragoon was cut across the elbow of his sword arm by a Russian horseman, at the heavy cavalry charge at Balaklava. The olecranon was completely detached, and the joint opened. The wound was immediately closed, the arm placed in an extended position, and cold employed to allay inflammation. Little more was done, and the divided surfaces quickly adhered, and an arm re-**mained which,** although not so free in its motions at that **joint as it was** formerly, was yet most useful, and would, I doubt not, become more so in time. Abscesses around the joint and œdematous swellings of the hand and arm are very apt to follow injuries of the elbow. Larrey thought gunshot wounds of the elbow particularly dangerous, from the strong ligaments which surround it, and the little distensibility of the joint; and he recommends amputation **when it has** been largely opened, even by a cutting instrument, **and blood has** been effused into the cavity. I have seen several cases in which, after being traversed by a ball, attempts have been made to save the elbow without excising it, but such trials were anything but encouraging. The motion of the joint, and its consequent use, will be found much greater after excision than when the arm has been saved without such an operation. Dupuytren has pointed out how important in gunshot wounds of the elbow the position of the aperture is. If at its inner aspect, the secretions get easy exit when the limb is in its natural position, and thus the chance of a favorable result is greater; if, on the contrary, the orifice be on the outer aspect of the articulation, no position will allow of the free flow of the pus except one, which will prove very fatiguing, and almost impossible to maintain for a long period.

The Returns show the following results as having been obtained from the resection of joints, from the 1st April, 1855, till the end of the war. The imperfect state of the official documents makes accuracy impossible with regard to the earlier part of the campaign :—

Cases.

Head of femur........................ 5, primary, of which 1 recovered.

"   "   .... 1, secondary, fatal.

Knee-joint............................. 1, secondary, fatal.

Os calcis, and part of astragalus. 1, recovered.

Os calcis alone....................... 1, recovered.

Head of humerus.................... 8, primary cases, 1 death.*

---

* Larrey performed excision of the shoulder in Egypt 10 times; 4 died—2 of scorbutus, 1 of hospital fever, and 1 of pest, after recovery. In 1795 Percy mentioned 19 cures after excision of the shoulder. Baudens had 13 recoveries from 14 operations, (*Rev. Med. de Chir.*, March, 1855.) Of 19 operations performed in Sleswick 7 were fatal, most of them from pyæmia. Legouest had 6 **cases of** primary resection of the shoulder in the hospital at **Const**antinople, of which 2 recovered. Thus, then, Hennen showed little discrimination in condemning the operation, when he says that it was "more imposing in the closet than applicable to the field."

Cases.

Head of humerus .... 5, secondary, no death.

"   and part of scapula .... 1, secondary case, followed by death.

Elbow-joint ....13, primary, with three deaths.

"   "   .... 4, secondary, died from causes not
connected with the operation.

Partial of elbow-joint............. 3, no death.

The above lists by no means represent the whole number
operated on. Those who underwent operation after the
Alma and Inkerman, after the battle of Balaklava, and the
first winter's work in the trenches, are all excluded, and
thus a vast number of the operations of the early part of
the war are omitted. In fact, I cannot but think that in
this way the *majority* of the operations do not appear, as
the number performed after these early engagements must
have exceeded those executed at a subsequent period.

The excisions of articulations injured by balls, although
occasionally performed during the Peninsular war, never became a very general practice, nor was it applied to some of
the joints which later years have shown its advantages in.
The Sleswick-Holstein campaign was the first great war in
which this conservative proceeding was followed out on an
extensive scale; and the results obtained by Langenbeck
and Stromeyer attest its efficiency, although they appear to
have had recourse to it in cases of very slight injury. These
operations certainly mark the surgery of the age, as, in the
words of Malgaigne, it may be said : " C'est une des plus
heureuses tendances de la chirurgie de ce siècle, quand la
necessité lui met le couteau à la main de ne lui concéder que
ce quélle ne peut lui ravir, de sacrifier aussi peu, et de conserver autant que possible."

The Crimean war afforded a considerable number **of** cases
adapted to the performance of **resection**; and I think our
results will stand a fair comparison with others, when all the
circumstances are taken into account. I will not say that
excision was performed in as large a proportion of cases of
injury of the joints as we were led to hope at the outset **of**

the war it might be; but when a better acquaintance was
had with the character of the wounds which this war presented to us, it was easy to understand how such should be
the case. The shafts of the bones leading from the joints
were often too extensively destroyed to enable the injured
parts to be removed by excision; in fact, the shafts were so
often split, and their periosteal and medullary membranes
destroyed, that the resection of the articulation did not suffice to save the limb. Surgeons soon recognized this; but
yet it was by no means always easy to determine the true
state of things about the joint till the incisions necessary
for resection laid bare the bones, and forced the reluctant
operator to convert his operation into one of amputation.

The great success which has attended the excision of
joints in civil practice, and a consideration of the fact that
the cases which fall to be operated on in the field are free
not only of local affections in the articulation, but also of
any active constitutional disease, made us all naturally sanguine of obtaining the best results from such operations in
the field; but unfortunately the circumstances above referred
to interfered so as frequently to leave us no alternative but
amputation.

So far as my observation went, primary excisions were
much more successful than those done at a late date; and
this fact is evidenced both as regards the final results and
the length of the period of convalescence, so far as we have
returns.

One advantage, with regard to the opportunity of performing resections early, is that we can in general tell at
first sight that such an operation *at least* is called for. It
is not so with regard to a large number of cases for amputation; but generally we can see at once that excision, at
any rate, must be performed, although we may not be able
to determine that subsequent amputation will not be required.

In the following observations I have not referred to the

amount of mobility retained by the joint after our excisions, as the patients went from under my notice too soon for my being able **to do so**.   It is certainly among the disadvantages of military practice during war that one can seldom trace cases to a conclusion ; not only so, but, in such cases as those of resected joints, very much of the after-result depends on that careful attention which no one can render so scrupulously as the operator himself ; and as it is never in his power to bestow this, we are not likely to have such favorable results as when, in civil practice, the patient remains, till finally cured, under the same hand.

The **shoulder-joint** is certainly that to which resection is most peculiarly applicable, from its superficial position and simple construction.   Interference with this articulation is, therefore, less disturbing to the constitution of the patient, and the results of the operation are more satisfactory, than those which follow a like interference with any of the other large joints.

The experience of the surgeons in Sleswick-Holstein led them to conclude that secondary operation is less unfavorable when the shoulder-joint is implicated than when a late operation is performed on any other joint.   The only secondary excisions of the shoulder-joint which appear in the Crimean returns seem to have been successful.

I know of a few cases in which what may be termed partial resections of the shoulder were performed, *i.e.* cases in which less or more of the head of the humerus was extracted, without the whole being removed ; and I believe the result of such cases to have been, on the whole, satisfactory, so far, at least, as the healing of the wound was concerned.   It is most curious how much can be done in this way.   The after-mobility of the articulation will, however, be more restricted in these partial excisions than if the whole joint is resected, and thus an entirely new joint-formation permitted.   The instances were not, however, very numerous in which the destruction of the bone was so limited as to allow of this

partial resection. Very much depends on the careful management of these cases afterward, especially in guarding
against inflammation.

In the old war, they restricted excision to those cases in
which the injury was confined to the **head of the bone**, holding that, when the shaft was much implicated, exarticulation
should be preferred, and this was very much the doctrine
acted on during the Crimean war. Guthrie thought the
insertion of the deltoid the lowest point at which the bone
could be divided with any prospect of success; but Esmarch
has shown that as much as four and a half inches may be
removed from the humerus, and yet a most useful arm remain. The ligamentous matter necessary to produce such
a favorable result requires a "plasticité" of constitution
which our patients did not possess.

The fact so clearly brought out by Stromeyer should be
always borne in mind, in determining on operations at the
shoulder-joint, that in comminution of the shaft of **a long**
bone, the fissures never extend into the epiphysis; in the
same manner, injuries of the epiphysis only in extremely
rare cases extend into the shaft unless the bullet strikes the
adjoining borders of both parts, in which case both are usually more or less seriously comminuted.

As to the best method of proceeding for resecting the
head of the humerus, some little difference exists, as is the
case with regard to the excision of some of the other joints
also, between those methods adapted for military and civil
practice. This arises, in a great measure, from the character
of the injuries necessitating the operation in either case.
The soft parts suffer little, and the bone is not diseased,
although broken, in cases operated on in the field, nor **are**
the parts bound together, as they so often are in **the** excisions performed in civil hospitals; hence it follows that we
can often remove all that is necessary through a much more
limited incision of the soft parts, than we could if disease
was the cause of the operation, or yet many of the accidents

which occur in civil life, in which, although the joint may be
but slightly implicated, the soft parts are yet so often greatly
destroyed. In the case of gunshot wounds, too, the periosteum, as well as the ligamentous and muscular tissues of
the articulation, can be retained, and thus a very great
advantage is secured, according to the views of this operation set forth by Stromeyer and Baudens. A single straight
incision will thus then very often suffice in resecting either
the shoulder or elbow joint, and even in similar operations
on the knee and hip, so that the maxim of Desault, that
"the simplicity of an operation is the measure of its perfection," is perhaps better exemplified in military than in civil
practice.   However, as in gunshot wounds two apertures
commonly exist, and as it is desirable to include them if
possible in the incision, we have a further illustration of the
saying, that the surgeon should be bound to no particular
form of operation, but should adapt his proceedings to his
case.

One of the chief dangers following resection of the shoulder
is the formation of sinuses and abscesses in the neighborhood. The best mode of avoiding this, is to arrange the
line of incision so as to give free exit to the pus.   Stromeyer's semicircular incision over the posterior surface of
the articulation fulfills this end better perhaps than any
other. The joint is there very easily got at.   Langenbeck's
one straight incision on the anterior aspect of the articulation, with or without the transverse cut suggested by Franke,
gave much satisfaction in Sleswick, where it was largely put
in practice; and I have myself seen most admirable results
got by the straight incision of White through the deltoid:
but Stromeyer's allows of the more rapid discharge of all
secretions than any of the others. Baudens, as is well
known, prefers a straight incision on the inside of the joint
in front, and from his large practice of it in Africa was
highly pleased with its efficiency, believing that it best
allows of that ginglymoid joint being formed, which, he says,

takes the place of the former articulation. Whichever of these methods of operation we adopt—and they represent those which have received the preference from military surgeons during late wars—it seems conceded by them all, that we do not require in field practice such extensive incisions as we do in civil life, and that by such limited incision the muscular and tendinous parts can be more respected, and thus the hope of restored action be much increased. The report of Esmarch, on the practice of Stromeyer and Franke, shows us that to cut across the fibers of the deltoid does not much interfere with its after-usefulness, "as its upper edge applied itself to and united with the articular surface of the scapula, and was thus fully attached and able to raise the arm. The healing was also quicker, as the space to be filled by granulation was much diminished in size by the application of the muscle to the glenoid fossa."

When the **neck of the bone** is broken here, as in the hip, so that it is difficult to seize the round head of the bone, the powerful forceps used by Mr. Ferguson in **excision of** the jaw, will be found to do good service.

Resections of the **elbow-joint** were more numerous in our army than in the French, yet the number of cases adapted for it in either force was but small. The numbers mentioned in the returns by no means, however, include all the operations of this kind which were performed.*

The formation of the elbow makes gunshot injuries of it much more serious than those which implicate the shoulder. Larrey was particularly gloomy in his prognosis of wounds of the elbow, and reports many disasters from them. That resection of the elbow is much less fatal then amputation, does not call for proof now-a-days, as it has been a **long-**

---

* Dr. John Thomson seems not to have been sanguine of the results to be obtained from excision as applied to either the shoulder or elbow. Of the latter he says: "I am satisfied that the difficulty of the operation, and the great length of time and care necessary for the cure, must prevent its adoption in military practice."

established fact.   The question now is more as to the extent
of the articular ends of the bones which can be removed,
consistently with retaining a useful joint.   My notes of cases
occurring in the Crimea, unfortunately, do not enable me to
throw any light on these interesting points.   In Sleswick,
out of forty excisions, only six patients died, and two others
were unsuccessful; but in thirty-two instances the effect was
very good.   "As regards two of them," says Esmarch, "I
have not been able to learn anything with reference to the
power of motion they possess; of the rest, eight have very
extensive, nine more or less complete, power of motion; it
is to be hoped of many of the remainder, that they will be
able to obtain much increased mobility by means of zealous
exercise of the arm.   On the other hand, thirteen of the
cases have a more or less complete anchylosis of the joint."

Several of the cases operated on in the East had undergone injury of all the bones entering into the joint, but no
case came under my notice in which so much as four or five
inches were removed, as was done in the war in the Duchies.

Partial resection—of which there were a good many cases
—did not, I think, turn out, on the whole, at all so well as
complete ones.   They were more tedious, more liable to
fail, and less satisfactory when they succeeded, than when
the whole articulation was removed.   The following were
cases of partial removal of the articulation:—

A soldier of the fusiliers had the head of the radius and a
small portion of the lesser sigmoid notch removed, shortly
after injury.   Amputation had to be performed two months
afterward, there never having been any attempt made to
heal by the parts.   A soldier of the 23d Regiment, admitted
into the general hospital in camp, after the assault on the
7th September, had the external condyle, the eminentia
capitata, and part of the trochlea, destroyed and removed,
the soft parts being little injured.   Sloughing set in, great
constitutional disturbance followed, and amputation had to
be performed three months afterward.   If complete excision

had been early performed in both of these cases, I believe
we might have obtained much more happy results. These
are only examples of several similar cases. But, on the
other hand, such cases as the following have occurred: A
soldier, when mounting the heights of Alma, was struck by
a rifle-ball, which passed across his elbow posteriorly, fracturing the heads of the radius and ulna, but leaving the
humerus entire. The broken fragments were removed, and
the humerus left untouched; and, after three months' careful
treatment, this patient was discharged with a famous joint,
which admitted of a considerable latitude of motion, and
with which he could sustain no small weight In the 9th
Regiment, a man was struck by a ball, which destroyed the
inner condyle of the humerus, without injuring the ulnar
nerve. The broken fragments were removed by Mr. Thorn-
ton, surgeon of the regiment. The subsequent inflammation
was commanded, and an arm was retained, which came
ultimately to possess three-fourths of its original motion.
Esmarch thinks that "the extensive severing of the ligamentous apparatus of the joint is what deprives the wound
of its danger, and that the less there is removed from the
joint ends of the bones, the greater is the probability of
anchylosis."

The complete fixture of the joint during the early period
of treatment, as so strongly dwelt upon by Stromeyer; its
constant support by a splint, even when being dressed; the
elevation of it, so as to prevent œdema; its flexure at an
angle of 130° to 140°, are all points of importance, both
as regards the comfort of the patient and the after-results.
Early passive motion before the wound is wholly cicatrized,
but at once abandoned if any irritation or signs of inflammation appear, are also indications which late experience
has stamped the value of.

The **hip** was **resected** six times; five being primary, and
one a secondary operation. One of the primary alone succeeded. Such success, although small, is yet encouraging,

as compared with the results obtained from amputation at
the hip, for which operation the excisions were substituted;
for, as will be afterward seen, out of at least 10 amputations
at this joint in our army, and some 13 among the French,
none survived, and in our cases at any rate the fatal result
very rapidly followed the operation.

I performed the first operation of excision of the hip
undertaken in the East, on the 6th July, 1855, on a rifleman,
whose case is subjoined :—

Couch, a soldier of the Rifle Brigade, was struck, on the
18th of June, by a ball, close below the elbow-joint of his
left arm. The ulna was fractured by the bullet, which then
struck the femur on the great trochanter of the same side.
The trochanter and neck of the bone was split, and otherwise severely injured. The patient did not come under my
care till the 5th of July, when I found a large, ragged wound
over the injured trochanter, from which a very profuse discharge of pus flowed. At the bottom of this wound the
bone was seen to be hollowed out into a large cavity, and to
be split in all directions. The bone was black and dead.
The limb was not shortened or distorted. The wound on
his arm gave him much annoyance, and the pain from the
hip was so great that he urgently requested some operation
to be performed, which might relieve him. He was brought
under the influence of chloroform, in order that the injury
might be more completely examined than could otherwise be
accomplished from the patient's irritable condition, and also
to enable me to get the limb put into a proper apparatus.
On enlarging the external wound, so as to make it dependent,
and to allow the necessary steps to be taken for the removal
of the dead portions of bone, a large fragment of the external part of the femur, which comprised what remained
of the great trochanter, was found detached, and a fissure
running upward, apparently into the capsule. It was found
impossible to remove the dead bone without opening the
joint; and, as but a very thin shell of the shaft was sound,

a consultation decided on the propriety of excising the head of the bone, and removing along with it what osseous substance was destroyed. This was done without difficulty, the original wound being increased a little upward. Hardly a drop of blood was lost. The wound was lightly dressed, and the limb fixed on an inclined plane, so arranged that the large dependent opening **retained** could **be** got at without having to remove the patient. The relief from pain and irritation which was experienced almost immediately was very marked and gratifying. Next day the patient's pulse **was** firmer, his expression very markedly better, and he declared himself as perfectly at ease. The limb, in a few days, was shortened about two inches. Suppuration became established; his strength improved; hectic disappeared; he slept well; and his appetite, which, before the operation, was nearly gone, was now restored, and he was totally free from pain or uneasiness. His pulse, on an average, continued about 85 beats in the minute, and was of good character. He continued to progress most favorably for about a week. Part of **the wound closed, and the rest of** it was **clean** and healthy. **At the end of that period** he **was** suddenly seized with violent diarrhœa, accompanied by vomiting and severe cramps, and followed by suppression of urine, which continued for 18 hours previous to death. His stools **soon** assumed the characteristic appearance of cholera evacuations, his strength sunk, he became rapidly collapsed, cold, and blue, and died during the night. Cholera was prevalent in the camp at the time.

After death, some crude tubercles were found in both lungs. There was no symptom of purulent absorption anywhere. The vessels in the neighborhood of the wound were healthy. There were old ulcerations in the intestine, and recent enlargement of the **solitary glands. The left** ulna was fractured obliquely up toward the elbow-joint. The wound over the hip was sloughy, an action it rapidly took on shortly previous to death, and the cut end of the bone was smooth and unchanged. **If I had** seen this patient **earlier, when the nature** of **the injury could** have been **more** exactly determined, I would have contented myself with merely gouging out the destroyed portion of bone, trusting to the remaining scale to throw out callus, fixing **the** limb carefully, giving free exit to the suppuration, an dstrenuously supporting the patient's strength. The state of the lungs and intestine, as revealed on post-mortem examination, made this patient, at best, but an unpromising subject for operation; but if the fatal disease which terminated his existence **had not** supervened, I would have been sanguine of the result. **When** he came under my care, I feel sure he was in **such a condition** that, if no operation had been undertaken to relieve him from the mass of dead bone which his system **was** futilely trying **to** get rid of, and which was setting up **further** disease all **around** it, **he would,** in a very few days, have **died,** exhausted **by** suffering. These are, however, points more easily settled on paper **than at the bedside.**

Mr. Blenkins, of the Guards, operated on the next case, and has been good enough to send me the following notes of it :—

"Private Charles Monsterey, aged 24, third battalion Grenadier Guards. Brought from the trenches at midnight with a severe shell wound on the outer side of the right **thigh.** Examination showed the thigh-bone to be extensively fractured at the upper part, in the situation of the trochanters and neck; the fragments were much comminuted **and the** surrounding muscles greatly lacerated. It was at once **recognized as an** appropriate case for excision, and the operation was performed half an hour after his arrival in camp. The wound was extended in a longitudinal direction to the extent nearly of five inches, and the shaft of the femur **sawn** through at the junction of the upper fifth with the rest **of the shaft.** The muscles were next detached from the trochanter, **and** the capsule lastly divided. It was intended, **at** first, to preserve the head of the bone in the socket; but the capsule was so extensively lacerated, and the cavity being filled with blood, it was resolved to remove it. Very little blood was lost during the operation. Examination afterward of the excised bone showed it to be fractured in fourteen pieces. The trochanter minor formed three, trochanter major three, shaft five, the neck three, besides numerous smaller fragments. The case continued to do well for the first three weeks; healthy granulations sprang up, both from the end of the divided shaft and the surrounding cavity and acetabulum. At this period pain and swelling of the knee-joint of the same limb supervened, the capsule of that joint became filled with purulent matter, the cartilages eroded, and he sank gradually, worn out with hectic symptoms, at the end of the fifth week, in spite of every effort to support him. The case was doing remarkably well, and I had every hope of his recovery until empyema came on."

Staff-Surgeon Crerar operated on the third case, a private of the first battalion of the Royals, who was wounded in the Greenhill trenches at mid-day on the 6th of August. The wound, in this case, was slightly posterior to the great trochanter, and was not larger than a shilling. It had been caused by a piece of shell, which, before it entered his thigh, had first struck a water canteen that hung by his hip. A comminuted fracture high up was clearly ascertained; but its exact position or extent was not defined previous to operation, although it was supposed to implicate the head and neck of the bone. "The trochanter was found broken into several portions, detached and imbedded in the contused muscles around, from which they were at once removed. The fracture was found to extend obliquely inward about an inch and a half along the shaft of the bone. The femur was now protruded through the wound, and I sawed off the whole of the fractured bone, leaving a smooth, clean surface; I then proceeded to disarticulate the head of the femur, which was effected without difficulty. Scarcely three ounces of blood were lost, and little or no shock was induced." This patient

26*

was seized with rigors, and died of exhaustion on the night of the 21st, *i.e.* on the fifteenth day from being wounded. The internal viscera do **not seem** to have been examined; but, **as** to the state of the femur, Dr. Crerar **says** "nature had not made the slightest attempt to repair the loss."

The **next** case, which was the only one where **success** followed the operation, was that of private Thomas Mackenena of the 68th Light Infantry, operated on by Mr. O'Leary, surgeon of **that regiment. The age of the** patient was 25; and he **was wounded** on the 19th of August, by a fragment of shell **which** struck him over the great trochanter and fractured it. It was thought that the fracture ran into the joint—a supposition which was disproved at the operation, **as the** head **of** the bone was uninjured. Five inches were **in all** removed. After operation, the limb was slung to the **beam of the hut. This** patient recovered in three months. I have lately heard of this man through the kindness of Lieut.-Col. Stuart, commanding the pensioners in **the Newry** district. Dr. Shaw, who signs the report, states that "the limb is two inches shorter than the corresponding one, and also considerably smaller; extension can be carried on par**tially,** but he cannot flex the limb upon the thigh without placing his hand on the glutei muscles of the diseased side. Rotation, inward and outward, can be performed only to a limited extent. The wound over the joint is quite healed. The man's general health is good, but he cannot walk without the assistance of crutches."

Dr. Hyde operated **on** another case, after the taking of the fortress. His patient, a private of the 41st, had the neck of the bone severely comminuted by a grape-shot, and died on the fifth day after operation. The cause of death is not given, nor can I discover it from the report of the case.

Dr. Combe, of the Royal Artillery, performed the sixth and last operation, on account of a gunshot wound of the **neck of the femur, in** which, however, the head of the bone was not implicated. This operation was not a primary one, but the patient survived a fortnight, and **died** of exhaustion, the most marked feature in the case being that the pulse remained very high—never below 120—during the period he lived, while his **aspect was** calm, and such as "might have led one to expect a more subdued state of the circulation."

Death thus followed in 2 (1?) from exhaustion, in **1** (2?) from pyæmia, **in** 1 from cholera, and in 1 from a cause that is unknown.

It is a remarkable fact that in these cases the head **of the** bone so often escaped, when the upper part of the shaft was fractured in pieces, which is probably to be accounted for by its protected position, and perhaps **by** the same cause as that **before** mentioned with reference to the head of the hu**merus,** viz., the non-extension of fracture of the shaft to the epiphysis.

Boyer and others have dwelt upon the depth of the parts, the strong ligaments, the difficulty of turning out **the head** of the bone, etc., as insuperable objections to this **operation ;** but actual experience—both in civil practice, where it has recently been so often **performed** for disease, and especially in our attempts in the Crimea, where the soft parts were in their natural condition, and the difficulty of turning out the head of the bone increased by the broken state **of the shaft** —proves that there is no such barriers to its easy execution. The greatest difficulty lies **in the** after-treatment. It is de**sirable to fix the** parts thoroughly, and at the same time to **allow** of some change of position. Mr. O'Leary managed this to some extent by means of a canvas sling for the limb. The fixture cannot, however, be satisfactorily accomplished in this way, whatever power it gives of changing **the posi**tion of the patient. I adopted the inclined **plane in** preference to the long splint, because **I** believe the position to be an easier one for such a case, and also because it permitted the free discharge of the pus and the easy dressing of the wound without disturbing the patient. If the idea lately

suggested at King's College, of slinging the whole body,
could be carried out, it might afford many advantages in the
management of excisions of the hip. As to keeping the
limb in a good position during cure, I fear more important
ends are lost sight of in striving after it. The uneasiness
and irritation which the splints and rollers give do much to
prevent success. It matters little what the resulting length
of the limb proves to be, if the patient's life is saved ; nor
does it greatly matter that it be somewhat out of the right
axis.

As to the comparative advantages of amputation and
excision at the hip in cases of compound fractures of the
head and neck of the femur by gunshot, some hint may be
got from our experience in the Crimea. Out of twenty-
three cases of amputation which took place, either in our
army or in that of the French, not one recovered ; and
nearly all died miserably, very shortly after operation. All
those, on the other hand, on whom excision was practiced,
lived in comparative comfort, all without pain, for a considerable time. Out of six operated on one survived for
more than a month, one died from causes unconnected with
the operation, and one case recovered entirely. The *chance
of saving life* is thus manifestly on the side of excision, and
this is truly the most important aspect of the question. The
objection so often advanced to the operation, that the limb
resulting from excision is useless, even if true, has nothing
to do with the matter. It is a question of deeper and more
serious bearing than such an objection would imply. The
only point worthy of discussion is, which operation holds
out the best chance of preserving life ? The little light derived from our Crimean experience is quite conclusive, so
far as it goes. In the one case a life was saved, while, out
of four times as many cases of the other operation, not one
survived. It is true that many cases submitted to amputation may have undergone more extensive injury than any
of those excised, and it is also true that one case of exarticu-

lation did, **to** all intents and purposes, recover; yet the shock of excision must be much less than that of amputa-**tion,** seeing that the great vessels and nerves are not touched, and that those changes **in** the blood of the limb are not interrupted, which some authorities contend is the cause of death **after** amputation. In all the cases of excision the loss of blood was trifling—a matter of much moment with **patients like** ours—and the immediate relief from pain and irritation was very marked in all the cases.

Gunshot wounds of the hip-joint are in many instances particularly adapted for resection, the injury of the bone being often limited, and the soft parts but little destroyed. There are, on the other hand, few accidents which present these conditions in civil life. When the shaft of the femur is split below the trochanter major, excision is hardly applicable, although Seutin performed it when he had six inches of the shaft to remove.

Seutin, Oppenheim, and Schwartz have all excised the hip **for** gunshot injuries, but not with success, although **all** seem to have been impressed **with its feasibility.** Paillard gives an account of Seutin's case, from which it would appear that the patient sank on the ninth day from gangrene. Six inches of the bone were removed in this case. In Oppenheim's case, the bone was removed as low down as the little trochanter, and the patient lived eighteen days. Esmarch relates Schwartz's case. **It was** a secondary operation. The bone, "to two inches below the small trochanter," was removed. He died of pyæmia on the seventh day after operation. Esmarch refers to another case, operated on by Dr. Ross, and related in the forty-first number of the *Deutsches Klinik,* 1850, which ended similarly. **This last** operation was performed two years after injury.

For **disease,** excision of the head **of the** femur has been now often performed,* and many times with success; the

* Dr. Sayre, in the *New York Journal of Medicine*, January, 1855, after relating **a** case of excision for morbus coxæ, gives a **summary**

very great difference, however, that exists between the operation as performed for disease and for accident, prevents any comparison being made between their results.

Much might have been done, if we had had another campaign, to determine the exact value of excision, **as applied** to gunshot wounds of the hip-joint. If the cases were selected with care, and the operation early performed, before the vital powers began to flag, and if the after-treatment were carefully conducted, much might be expected from this operation in military practice. It is often very difficult to tell how far the destruction of the bone extends, either upward or downward; but if the case should turn out to be too complicated for excision, then amputation may be **performed.** Stromeyer has shown that, although the splitting of the bone barely extends into the capsule—as it did in my case—yet excision should be at once performed, as suppuration **is sure to be set** up **in the** articulation, and death by exhaustion follow. The same surgeon **has also** shown how it happens that, although **the neck of the** bone be fractured by a ball, yet the usual signs of such an injury—the shortening and rotation of the foot—may be absent, from "the fragments hanging together better on account of the partial preservation of their fibrous covering," and, in one case which he examined, a considerable power of flexion and extension remained, although the neck of the bone was fractured; while, in another case, "the fragments fitted so well together that the patient did not experience the least pain, and the leg could be moved without causing crepitation." The existence of the fracture was only determined in this case by the presence of a profuse discharge. The patient himself may even be able to

---

of the operations of this sort performed up to that time. I need hardly add, that many have been performed since Dr. Sayre wrote. He classifies 30 cases, of whom 20 recovered and 10 died, only 4, however, within **a week of** the operation.

move his foot, and so mask the diagnosis. Esmarch gives "the extensive swelling occurring rapidly, and the pain on motion," as the only two signs which are nearly always present.

It need hardly be added, that if, in fracturing the **neck of the bone**, the ball or any of the osseous fragments injure the great blood-vessels, the case is not one for excision.

The **knee** was only once excised, so far as I know, during the war. The operation was performed in the general hospital by Mr. Lakin, whose notes of the case have been kindly furnished to me.

"Henry Gribben, aged 19, a private in the 77th Regiment, was admitted into the general hospital on September 8th, 1855. While retreating from the Redan on that day, he received a musket-ball in his left popliteal space, causing him much difficulty in walking; nevertheless, he succeeded in regaining the advanced trenches, distant about 100 yards. He was a man of average muscular development, and of habitually good health. On admission, a circular wound, with inverted edges, was found at the inner part of the popliteal space, and at the level of the junction of the upper and middle thirds of that space. It was of a diameter just sufficient to admit the index finger, which could be passed to its full length in a direction forward, and slightly upward, between the inner hamstring tendons. No fracture nor other injury of the bone was detected, neither could the further course of the ball be ascertained by means of a probe or elastic catheter. It was not considered prudent to use much force with these instruments, in consequence of the close proximity to the joint, and of the absence of any satisfactory evidence that its cavity was already opened. There was no aperture of exit, the limb was not altered in shape; flexion and extension, especially the former, **were** limited, and any attempt to move the limb beyond these limits was attended with much pain, which was otherwise slight. Simply bearing the weight of the body only caused some uneasiness, and

there was no tenderness on pressure from without. There
was no appearance of synovia about the wound, nor was
there any bleeding.   Under these circumstances it was con.
sidered that any operative measures for the purpose of
removing the ball were not justifiable."

The limb was placed upon its outer side, with the knee
semiflexed, that being found the most agreeable posture, and
cold dressing was applied. The patient remained almost
free from pain, except when the limb was moved, and in
good health, until September 20th, twelve days after the
injury, when the joint became somewhat inflamed, as indicated by increased pain and heat, slight tenderness to
pressure, and moderate swelling. Twelve leeches and hot
fomentations were applied, and afforded great relief. The
symptoms subsided, and remained in abeyance till about the
29th, when they gradually increased, the joint becoming
much swollen and tender, the veins more distinctly visible,
and the general health beginning to suffer for the first time,
as evinced by slight perspirations, debility, frequent pulse,
loss of appetite, thirst, disturbed rest, etc. The swelling of
the joint was uniform, and no fluctuation could be perceived
in it, though it was thought that there was some deep-seated
fluctuation about three inches above the joint, on the outer
side of the thigh.

"It was decided in consultation to examine the limb while
the patient was under the influence of chloroform, and then
to adopt such measures as the examination might indicate;
accordingly, on October 1st, he was placed upon the operating table, and chloroform administered. With some difficulty, and by using considerable force, the finger could feel
a part of the head of the tibia, bare and rough, a small piece
of bone having been chipped off its inner and posterior edge,
but the site of the ball was not detected, though it was
thought to be in the joint, possibly in the space between the
condyles of the femur. It was then decided to make such
an incision as would admit of the performance of either ex-

cision or amputation, whichever proceeding the condition of
the parts might indicate. This was accordingly done; and
on opening the joint several portions of the cartilage covering both bones were found to be partially detached from the
bone, softened, and their surface eroded. No fracture was
found, except the small piece chipped off the inner and
posterior edge of the head of the tibia."

"Excision being now decided on, and as the necessary
steps were being taken, pus escaped from a cavity which
existed in the outer side of the thigh, and partially surrounding the femur. The ball was found to have penetrated the
inner condyle. About an inch and three-quarters of the
femur was removed, as well as a thin slice of the head of the
tibia. The patella was also dissected out, because portions
of its cartilage were softened, and partially detached. The
slight oozing of blood was soon stopped by cold water."
No vessel required a ligature. The edges of the wound
were brought together, and retained by sutures and strap-
ping. The extremities of the incision were left open, to
allow of the escape of pus, etc. Wet lint was applied, and
the limb placed in a straight position on a M'Intyre's splint,
with a short whalebone splint on each side of the joint,
secured by strap and buckles. The patient was placed in
bed, and a grain of morphia given him.

"The portion of femur removed was about one and three-
quarter inches long, and presented an ordinary round musket-
bullet, about half imbedded in the inner condyle, the bone
not being split, but the joint opened."

No symptom arose calling for remark up to the 25th,
when Mr. Lakin's report runs thus: "Had continued slowly
improving and gaining strength until to-day; the discharge
had diminished in quantity. Had not accumulated nor
bagged. The limb had acquired slight firmness. The
wound looked healthy, and had nearly healed across the
front. Some difficulty had been found in keeping it in very
accurate position, as he twisted about when using the bed-

27

pan, and he is naturally a reckless, troublesome fellow. His
bowels were occasionally slightly relaxed, but this was soon
relieved by a dose of the aromatic mixture. To-day he seems
progressing favorably, but has got his limb into a bad position; bent so as to form an angle externally. A slight discolored patch, as of a commencing slough, on the outer side
of the limb, corresponding to the position of the displaced
end of the femur, at the upper extremity of the wound.
The plane is readjusted, and the limb secured to it by bandages. The discharge is again rather increased in quantity.
A bad sore had formed upon the sacrum, but is improving
under treatment."

Again, on the 27th, the report says: "Complains rather
of chilliness this morning, but has had no rigors. Has
vomited several times, and his bowels have been purged.
Pulse 110. Tongue moist and clean, wound healthy, small
slough on outer side not extending, discharge us usual, urine
drawn by a catheter."

The diarrhœa, although temporarily checked by treatment,
went on, and the sickness greatly prostrated his strength.
Mr. Lakin notes as follows on the 28th: "Rapidly getting
worse. Pulse 130. Very low; evidently sinking; countenance much altered, but simply looking sunken and pale, and
not having the peculiar aspect of pyæmia. Died at night."

Post mortem 14 hours after death. "Before removing
the body to the 'dead tent,' the orderlies had taken off the
splint, and the limb had been allowed to hang down, so as
to destroy any points of union that there might have been.
The wound had healed, except its extremities, the granulations on which had shrunk and assumed a black appearance,
(post mortem;) the opposed surfaces of the bones presented
a very similar appearance, and there was no sign of dead
bone. They had become moulded to one another in shape.
Whether there had been any union toward the center was
not evident; at the circumference there were appearances of
some adhesions having been broken. The cavity of the joint

contained only a small quantity of pus. The abscess in the
outer part of the thigh had almost healed. No purulent
deposits could be found in any of the organs, nor could any
appearance of phlebitis be detected. The viscera were
healthy."

"I ascertained," adds Mr. Lakin, "after his death, that,
on the 26th and 27th, he had eaten some apples which he
had bought, and that the vomiting and diarrhœa came on
after that. He had not at all the appearance of a man suffering from pyæmia, but seemed simply to die exhausted by
sickness and diarrhœa."

"The opening through which the bullet entered remained
patent all the time, and a great deal of the discharge escaped
through it; though probably the two extremities of the incision would have been sufficiently on the posterior part of
the limb to prevent the matter from bagging."

Admiring as I do the brave attempts which have been
made in civil practice to save limbs by excising the knee, I
regret that it should not also be extended to military practice; but except in rare circumstances, I fear that cannot be
accomplished, from the careful after-treatment and the long
period of convalescence necessary to effect a cure. Ferguson
speaks of more than 100 cases having been now operated on
in civil practice, and Butcher has shown that the mortality
is greatly less than what succeeds amputation of the thigh;
but it is to be remembered that these cases were of an age
and a history which rendered the procedure much more hope-
ful than it almost ever can be in warfare. A diseased joint
is a constant source of irritation and depression to the constitution, so that, in the words of Sir Philip Crampton, "by
its total excision all those parts which were diseased, and
influenced the constitution **so** unfavorably, are removed from
the system, and the injury **is resolved into a** case of clean
incised wound, with a divided but not fractured or diseased
bone at the bottom of it," and thus the powers of the system
which went to feed the disease are already so diverted to the

part as to build up the loss, so soon as they can work on a proper material. That nice adaptation, however, of the surfaces, that accurate fixture of the limb, the careful attention, nourishment, and perfect repose which such cases obtain in a civil hospital, and which have so much to do with the result, can hardly be attained in the field. Mr. Ferguson, in the last edition of his admirable manual, thus sums up the advantages which his large experience ascribes to the operation: "The wound is less than in amputation of the thigh, the bleeding seldom requires more than one or two ligatures, the loss of substance is less, and probably on that account there is less shock to the system; the chances of secondary hemorrhage are scarcely worth notice, as the main artery is left untouched; there is, in short, nothing in the after-consequences more likely to endanger the patient's safety than after amputation, while the prospect of retaining a useful and substantial limb should encourage both patient and surgeon to this practice."

If the operation be performed in the field, the sooner it is undertaken the better; for, although primarily free of disease, the articulation soon becomes affected, if it be left a prey to inflammation and abscess; the constitution rapidly sympathizes, and that blood-poisoning which is so liable to follow may be established before we well see the danger of delay. Secondary operations, too, it should always be remembered, do not hold out the same prospect of success in military as they do in civil practice.

The saving of blood, and the absence of any fear of secondary hemorrhage which has been pointed out by Butcher and Ferguson, are points of much weight in favor of resection when patients are to be dealt with who are so sensitive to any hemorrhages as those we had to deal with in the Crimea.

The resection of parts of the shafts of the long bones was not, to my knowledge, much practiced in the Crimea. The lengthened period those cases take to recover, and the trying nature of this ordeal on the vital powers, made such abstinence with us almost a necessity. Several cases resulted very favorably, in which part of the shafts of the humerus, of the bones of the forearm, and of the leg, were thus **dealt** with; but in more than one case in which I knew such steps taken, too much was expected of the reparative powers of our patients, too large an extent of the **bone was** removed, and thus the operation failed. It was toward the **end of the** war that the best results were obtained from these resections. In the case of the tibia especially the choice between amputation and resection must be guided chiefly by a consideration of the state of health of the patient, whether or not he is in a condition to withstand a long and tedious cure; by the extent of destruction of the bone, and especially of its periosteum; and, finally, the means at hand for carrying **out** the after-treatment.

Resections in the continuity of the femur were, so **far** as I know, invariably fatal. The difficulty **of the operation** on muscular limbs must of itself predispose to disagreeable results.* **False joints are, as is well** known, apt to occur after resection in the continuity of bones of the leg and forearm, when the operation is practiced on only one of their two bones.

---

* The success of two cases lately published by Mr. Jones, **of Jer**sey, in which a considerable part **of the shaft of the femur** was removed, shows that these operations **may be well adopted in** civil practice at any rate.

27*

CHAPTER XII.

AMPUTATIONS.

THE relative advantage of **primary** and **secondary amputation** has always held the first place among the various problems which the army surgeon has had to solve. With all that has been written on the subject by military and civil surgeons, there still seems considerable reluctance to accept the question as settled. The discrepancy of evidence **brought** to **bear on the** subject has chiefly arisen from the evident distinction being overlooked between operations undertaken for accident and for disease. Civil hospitals can seldom afford testimony similar to that obtained from the field of battle, and thus it happens that civil surgeons have come to stand in some measure in apparent antagonism **to their** military brethren on the point of practise under consideration. Hunter **was** so much of the civilian as to adhere to consecutive operation, although, with very few **exceptions,** surgeons who have practiced in armies have **strongly** advocated early interference since the days **of Duchesne** and Wiseman.* The difference which **so mani-**

---

* Hunter surely does not express the opinion of military surgeons, with perhaps the exception of Faure, when he says: " I believe it is universally allowed by those whom we are to esteem the best judges, those who have had opportunities of making comparative observations on men who have been wounded in the same battle, some where amputation had been performed immediately, and others where it had been left till all circumstances favored the operation ; it has been found, I say, that few did well who had their limbs cut off on the field of battle, while a much greater proportion have done well, in similar cases, who were allowed to go on till the first inflam-

festly exists between the moral condition of the patients who
are operated on for accident in civil life, and the soldier in
the field, together with the circumstances in which each is
treated after operation, introduce so many different items
into the calculation of the question of amputation, that it is
almost impossible to make use of the experience of either
sphere to illustrate or influence that of the other. Besides
this, the severity of those injuries which present themselves
in military practice, and which authorize the removal of the
limb, is so great that it is but reasonable to suppose that
an operation which removes so vast a source of irritation
and pain at the earliest moment possible, must promise the
best results in saving the life. In short, military experience
on this point must regulate military practice, and the results
of civil experience must continue to regulate civil practice.

To military surgeons the question of primary or secondary amputation is a settled one. The experience of every
war has more and more confirmed the advantages of early
operation, and that in the Crimea has not disturbed the
rule; in fact, later observation would lead us to go further,
and in place of merely advocating interference within twenty-
four hours, the prevailing idea at present would be better
expressed by saying that every hour "the humane operation"
is delayed diminishes the chances of a favorable issue.

It is impossible to prove from any returns the full bearing
of this question, as the mere number who survived after a
given number of operations performed primarily or secondarily by no means expresses the terms of the question. It
would manifestly be necessary to know how many died

mation was over, and underwent amputation afterward " **Contrast**
such sentiments with the immense mass of facts scattered throughout
the works of Larrey, Guthrie, Hennen, **and, in fact, of** every military surgeon who has published **his** experience **in** this **or** foreign
countries for many years, and then **let us be** led, in estimating this
long-contested question, by what Hunter says is "the best guide,"
viz., experience.

before the secondary period came round, and to these should be added the victims of delayed interference, with all the pain and suffering which such delay occasioned, before we can arrive at a just estimate of the results of either proceeding. The experience in the Crimea in favor of early operation was unequivocal in both armies, and needs no illustration from me.*

Chloroform has done much to render the success of primary amputation, as contrasted with secondary, yet more marked. If we believe, as I certainly do, that by the use of this anesthetic all fear of intensifying the shock is obviated —which was one reason why surgeons delayed operation— then the tendency of military surgery, since the introduction of chloroform, must be to still earlier and more prompt interference.

Secondary amputations were much more common during the early than the late period of the war—a circumstance which arose from the deficient means of treating the wounded in the camp during the former as compared with the latter period, and thus the necessity that existed of dispatching them from camp immediately after being injured; and this, together with the better hygienic condition of the patients toward the end of the war, accounts for a fact, well known to those who served in the East, but which the range of the returns does not enable me to show in figures, that amputations were much more successful, as a whole, toward the conclusion than at the outset of the war. At first, too, when patients were early sent from camp, not a few operations, to my own knowledge, were performed during the "intermediary" period, and, without one exception, those thus falling within my observation were fatal.

The tremendous destruction which was at times occa-

---

* I am led to understand, from a very well-informed source, that the Russians also lost two-thirds of all their secondary operations, but saved a fair number of their primary.

sioned by round shot or shell left little hope from any operation whatever. In the case of many, a *pansement de consolation* was the only alternative, while, in not a few, the injury was so severe that, although amputation was performed, in the vain hope of a possible success, yet the apparent advantage of primary operation thereby suffered, and this circumstance is another of the many which makes it impossible to place this question in a fair light. The most severely injured have their limbs removed early, while the most hopeful are retained for secondary operation, and thus all the advantages of slighter injury—less constitutional disturbance, more promising habit of body, and state of general health — are denied to the early operations. In truth it may be said that if, with all the advantages under which secondary amputations are recorded, they appeared as merely equal in success to primary, then the superior claims of the latter to our attention would be sufficiently clear; how much more marked, then, are the successes of early operations, when we find them giving such superior results!

As to the general success of amputations, during the late war, it may be safely said that, when due weight is given to the many circumstances which have militated against the success of all operations, and which have been fully dwelt upon in the course of the preceding pages, those performed early have afforded a very fair proportion of success; while it cannot be denied but that those undertaken late have been followed by most unfortunate results.

A siege presents peculiarly favorable opportunities for testing the value of **immediate amputations**. The men being close together, and acting within a narrow space, can be seen almost instantly on being injured. The position of the soldier in such circumstances resembles that of the sailor on board of his ship; so that the experience of naval surgeons, which is so strongly in favor of instant amputation, applies with peculiar force to military siege practice.

Unfortunately, the arrangement followed in our army during
the siege of Sebastopol made the elucidation of this point
impossible.   Assistant-surgeons were alone sent to the
trenches, (except during an assault, when a staff-surgeon
occupied one of the ravines behind each division; but in the
hurry and confusion which prevailed at such times, the men
he operated on were lost sight of;) and as by the rules
which prevail in our service, an officer of that rank is not
allowed to amputate, except when the surgeon is not with
the regiment, no means existed for the due examination of
this question.   The French experience, if it were available,
would be of much use on this point, as they performed many
capital operations in their trench ambulances.

Whatever that condition is which is conventionally known
as 'shock,' it seems pretty evident, from the admission of
all, that it is not established for some little time after the
receipt of injury—an interval which differs in duration
mainly in accordance with the severity of the wound, the
agency by which the injury has been caused, and probably
the constitution of the sufferer.   The evidence of naval
surgeons, as summed up by Mr. Hutcheson, in reference to
the absence of shock immediately after the receipt of a
wound, must be conclusive to all unprejudiced minds; and
instances were not wanting during the late war which
appeared to support the same view.   I know of several
well-authenticated cases which occurred during the siege,
in which the perfect absence of all constitutional prostration
after an accident so severe as the carrying off of a limb, and
the non-appearance of such shock for some considerable
time after, went to prove the same position.   If this precious
moment could be seized at all times, and that operation
performed under chloroform, which assists so much in ward-
ing off the "embranlement" we fear, how much more successful would our results prove, than under any other
circumstances they ever can be!

It is during this interval, too, that we obtain the full

good of the soldier's *moral* advantages over the civilian. "Cut off the limb quickly," says Wiseman, "while the soldier is heated and in mettle"—and the observation is as old as Paré, that while excited by the combat, and yet within sound of the cannon, the soldier or sailor willingly parts with a limb which a few hours of reflection would make him desire to run the risk of preserving, and upon which he fixes all his attention, so as to magnify greatly the dangers of the subsequent operation. Moreover, the removal of the man, before operation, to any distance from the scene of his accident, lessens somewhat his chances of recovery; as, besides the danger that the irritation and pain of such transport, however carefully it may **be** conducted, will occasion—the constitutional depression we dread; the mere **loss** of blood which, although going on in very **small** quantities, is yet flowing in drops, when a drop may extinguish life, are serious objections to the shortest delay.

But even although that constitutional disturbance which **is** the result of injury is present, is it always necessary **to wait** its subsidence before operating? If it be very decidedly marked, and the patient thus much prostrated, such delay may certainly be called for; but it is an opinion often stated by those who must be well informed on the subject, that such delay is not always advantageous, but manifestly the reverse. Larrey, for example, gives repeated utterance to the following sentiment: "Il est donc demontré que la commotion, loin d'être une contre-indication à l'amputation primitive, doit y determiner le chirurgeon;" and again: "Les effets de la com- **motion loin de** s'aggraver, diminuent et disparaissent insensiblement apres l'operation;"* and in this opinion he is by no

* "J'ai vu perir un assez **grande nombre de militaires** quoique opérés dans les vingt-quatre heures **parce que l'operation** avait encore été faite trop tard. Sur trois **que j'ai amputés aux** deux jambes **à la** bataille de Wagram celui qui **avait été** opéré le premier et peu d'instans apres le coup, est le seul que ait été sauvé."—*Larrey*, vol. iii. p. 378.

means solitary, as may be seen by reference to the writings
of many naval surgeons, who have manifestly the best opportunities of judging in the matter.   The upholding influence
of chloroform comes strongly into play in such cases, and
obviates, in a great measure, the dangers which have been
prognosticated from such proceedings.   If the constitutional
depression be the result of an injury which remains as a
source of irritation, then the removal of such must manifestly be a great point gained; and I know it is the opinion
of many army surgeons of large experience, that the presence
of shock is no hinderance to operation, (under chloroform,) if
that condition be not very decidedly marked at the moment
of interference.

The difficulty which chiefly stands in the way of instant operation is the recognition of the cases which demand it, and
the certainty that no fatal internal lesion may not have been
at the same time sustained, as the accident to the limb which
necessitated its removal.   However, it would certainly tend,
on the whole, to the saving of life, to operate as soon as
possible, not only in all those cases in which the necessity for
it was evident, but also in all doubtful cases; as, although a
few limbs might thus be sacrificed, I have not the least doubt
but that many lives would be saved.*

* In fact, there can be little doubt but that it would tend greatly
to the preservation of life in an army on active service, if it were
made imperative on surgeons to operate in certain contingencies, in
place of leaving it, as at present, to the discretion of each what cases
to preserve and what to operate on; as the undoubted tendency
among surgeons—notwithstanding the prejudice which so long existed to the contrary—is to amputate too little on the field of battle.
I know full well such a regulation could not be made, nor would it
be withstood by a medical staff; but judged of merely as bearing on
results, I have little doubt but that it would be successful.   As was
said when speaking of compound fractures and wounded joints, every
succeeding generation of surgeons go through, to a great extent, the
same ordeal in gaining their experience.   They suppose their advanced attainments encourage an attempt which their predecessors

The Crimean war afforded a most excellent field for observing the relative value of flap and **circular amputations**; as, although in our army the former was commonly employed, most of the French and not a few of our own surgeons adhered to what Sir C. Bell termed "the perfection of the operation of amputation." As the advantage, in general, of removing the limb as far as possible from the **trunk is** fully recognized, it seems curious that the circular mode of operating, which I think admits of this more than the operation by flaps, should not be more followed. In the lower part of the thigh this is particularly observed. Protrusion of bone is the great bugbear which terrifies most operators; hence they make unnecessarily long flaps, and remove a much **larger amount of** the bone than is at all necessary. This **was very** apparent **in** many amputations in the East. Mr. **Syme has** laid down the true principle which should regulate our proceedings, when he says: "It is not the length of the flaps which prevents the risk of protrusion of **the bone,** but the height at which it is divided **above** the **angle of union** of the flaps."

In soldiers, **as in many** (although not all) cases submitted to primary amputation for accident at home, the proportion

---

feared, **and thus a vast number of lives are** ever being sacrificed to the establishment of individual experience. Were it possible that a **commission,** a chief, an academy, or any competent **body** having **authority,** were to lay down instructions at the beginning of a campaign binding on the surgeons of an army, with *reference to points fully established by a large and sufficient induction,* as well as those which called for their investigation, I have little doubt but that a large proportion of lives would be saved during each war, and **a mass of** reliable facts added to our knowledge. The fluctuating **state of our** knowledge upon those cases which **demand** immediate amputation might be thus thrown **into shape and made available,** while those operations would alone be **made** imperative **the justness of** which was beyond doubt. Such **an** arrangement as **that hinted** at, in place of being a hinderance to the advance **of** our knowledge, would, in truth, promote **it.**

of muscle to skin and subcutaneous fat is different from what
it is in most cases operated on in civil hospitals, and thus
modifies our appreciation to some extent of the two modes
of operating.   In soldiers there is commonly but little subcutaneous fat, and the muscles are large and strong; hence
it becomes very difficult, when practicing the flap operation,
to adapt the parts to one another, so as to fulfill the latter
part of the old maxim, "muscle must cover bone, and integument muscle."   It cannot be said that this arose in the
East from the maladroit performance of the operation by
the flap, as the same circumstance may be seen to occur at
home in the hands of our ablest hospital surgeons.   The
paring and stuffing-in processes which are not uncommonly
seen in hospitals, to correct the results of the condition
referred to, are no less prejudicial than unsightly.   The irritation is thereby increased, and proper adhesion of the parts
prevented.   In secondary amputation the excess of skin removes any fear of similar accidents.   Chloroform has refuted
the argument in favor of the flap operation, founded on the
greater speed of its performance than the circular, as such
great speed is now a matter of no moment.   But however
it be with regard to the question in general, there is one fact
which any one who had opportunities of watching matters
during the early part of the late war will amply verify, viz.,
that the circular stumps stood the transit to the rear much
better than those formed by the flap method, and thus it
would seem that the former mode of operating is more advantageous in military practice than the latter.   The long,
heavy flaps were so knocked about during the land and sea
passage that they often became loose, got bruised, and ended
by sloughing; while the firm, compact stumps made by the
circular method were little if at all injured.   When patients
can be treated in camp to a termination, the influence of
this circumstance is, of course, null.   It may be said that
the length of the flaps was a mistake committed in the operation, but, unfortunately, such errors must always be looked

for in like circumstances, where there is a large body of operators, most of them without previous experience in operating, and whose chief fear always is to have "too little flap;" for although it is true what Hammick says, that "it requires more practical experience to know when to take off a limb than how to do it," yet the *how* must also be studied, like everything else.

In considering the **statistics** of amputation performed during the Crimean war, the figures refer solely to the period between the 1st of April, 1855, and the end of the war, and consequently exclude all the unfavorable part of the campaign, as well as the greater number of the operations which were absolutely performed during the war. It was found impossible to attain to accuracy with regard to **the ear**lier period, so the field of observation was restricted as stated. It is needless to point out how different must be the lessons derivable from the statistics of this latter period alone, to what they would have been if the whole period of the war had been included.*

During the limited period I have mentioned, there were 732 amputations in all parts performed, followed by death in 201 instances; of these, 654 operations and 165 deaths were primary, and 78 operations with 36 deaths, secondary; giving a percentage of 27·4 deaths overhead—25·22 for the primary, and 46·1 for the secondary operations. If we include only the greater operations, viz., amputations of the shoulder, arm, and forearm, of the hip, thigh, knee, and leg, **then** we have a total of 500 cases and 199 deaths, or 39·8 **per** cent.; of which total 440 cases and 163 deaths, or 37

---

* In my original papers the figures were intended to represent **the** period of the whole **war**. I have reason **to think that,** although upon a more careful investigation **of the** returns than could be made in the Crimea, these numbers have since proved not strictly accurate, they yet represent pretty much the results which followed many of the operations as viewed in the more lengthened and less favorable aspect of the war.

per cent., were primary, and 60 cases and 36 deaths, or 60 per cent., were secondary.*

The increase of the mortality as we approach the trunk may be shown thus, taking the primary amputations alone as giving the most unbroken series :—

SUPERIOR EXTREMITY.

Part.	Ratio mortality per cent.
Fingers	0·5
Forearm and wrist	1·8
Arm	22·0
Shoulder joint	27·2

INFERIOR EXTREMITY.

Part.	Ratio mortality per cent.
Tarsus	14·2
Ankle-joint	22·2
Leg	30·3
Knee-joint	50·0
Thigh, lower third	50·0
" middle "	55·3
" upper "	86·8
Hip-joint	100·0

The lower extremity was removed at the hip-joint seven times during the period included in the returns, and at least three times more previously, giving ten cases, all primary operations, and all ending rapidly in death. One of these cases was operated on by my lamented friend Dr. Richard M'Kenzie, after the Alma. The French had thirteen cases, primary and secondary, after the Alma and Inkerman, and all died. One of these, a Russian, was operated on by M. Legouest on the 3d of October, 1855, at Constantinople. The upper part of femur was completely smashed by a conical ball. The flaps had adhered to a point by the middle of December, at which date I saw the patient walking about the ward on crutches, and looked upon by all as being be-

* See Appendix G.

yond danger. The very night on which the order arrived
for sending him to France—where he was to be admitted,
by special permission, into the Val de Grace—he fell when
walking in the corridor, and hurt his stump so that it bled
profusely. Inflammation was set up, suppuration, renewed
hemorrhage, and diarrhœa followed, and he died on the 9th
of February, four months after operation. M. Mounier, in
the same hospital, had three cases, one of which I watched
with interest. Two of these died of hemorrhage, one on
the fifteenth, and the other on the twentieth day. The third
died of cholera. One of these men was a Russian.

The mortality which has thus followed exarticulation at
the hip, during the Eastern campaign, has been very deplorable; yet, in the cases in which it was performed, no other
alternative remained, except to abandon them to inevitable
death, which many might be disposed to think the more
humane proceeding, as they often linger for a long period
before death. M. Legouest's case was unquestionably successful; and, although we can hardly hope with Larrey that
this operation will **ever** be performed as readily as **his**
favorite one at the shoulder-joint, still the results of operation at the hip for accident have not been so utterly hopeless as to lead us to abandon it. M. Legouest has given,
in a most interesting paper on the case mentioned above, a
table containing most of the recorded cases of amputation
at the hip for gunshot wounds. Of primary operations he
has collected 30 cases, all ending fatally; of intermediate or
early secondary operations he finds mention of 11 cases,
with **3** recoveries; and of operations performed at a period
so late as that "the injury had lost all its traumatic character," 3 cases, with one recovery. Thus, if **we sum up the**
whole, we have 4 recoveries in 44 **cases, or a mortality of**
90·9 per cent. Some of the primary cases died on the
table; all of them before ten days except 2, which perished
within a month. The proportion of recoveries among those
operated on after the primary period, but before a long

elapse of time, *i.e.* at some period during the existence of
"the traumatic phenomena," was the largest, and hence that
would seem the best time to undertake the operation.

During the Sleswick-Holstein war, **amputation at the hip**
was performed 7 times—5 were operated on by Langenbeck;
only 1 of these cases recovered. I find no mention whether
these cases were primary or secondary. In the Indian
campaigns I find mention of only 1 case of amputation at
the hip for a gunshot wound. It was a primary operation,
and took place in the Punjab. Thus, if we reckon the
whole number of cases operated on for gunshot wounds,
those recorded by Legouest, our own Crimean cases, and
the Holstein and Indian ones, we find a total of 62 cases,
and 5 recoveries, or a mortality of 91·9 per cent.

Mr. Sands Cox, recording the experience of civil hospitals as well as those of military practice, up to 1846, gives
in all 84 cases, most of them for injury, with 26 recoveries;
14 of these successful cases being after accident, and of the
unsuccessful, 20 were for injury; and in the *Medical Times
and Gazette* for April, 1857, there is a further record of 8
cases, of which 2 were for accidents, (1 primary and 1
secondary,) with 3 recoveries, all after operations for disease.* Cox recognizes the difficulty of restraining the
hemorrhage during the operation, and the shock given to
the nervous system, as the great sources of danger. The
hemorrhage, at a considerable period after operation, would
appear even a more common cause of the fatal event than
the difficulty of commanding it at the time.

---

* In the *New York Journal of Medicine* for October, 1852, there is
a paper by Dr. Smith, on the subject of amputation at the hip, in
which he gives a summary of 98 cases, showing a ratio of mortality
of 1 in 2⅔. In 62 of these cases, of which he learned the particulars,
the operation was performed for injury in 30 cases, and the percentage of deaths was 60. He remarks one curious circumstance,
viz., that the ratio of mortality has most suddenly and markedly
diminished since 1840, and no reason can be given for this, unless it
be increased care, better operative ability, *and the use of anæstheti s.*

It will, of course, only be in the event of such destruction to the bone or soft parts, or such other injury to the nutrition of the extremity, as puts resection out of our power, that amputation will be performed. If the fracture of the neck of the bone were slight, as when occasioned by a small ball, or one striking with little propulsive force, such as that projected by the match-lock, then the case, I conceive, must be viewed more as a compound fracture of the upper part of the thigh, and should be treated accordingly. M. Legouest has recorded 6 cases in which the limb was not removed or resected, and 3 of these recovered. One of these cases of recovery having occurred in 1812, **must** have been wounded by a round ball; the second was injured in a duel, and hence probably by a small light ball; while the third was observed in Africa, where neither the size nor the form of the balls **used** by the natives is to be compared to the conical bullet. All three were struck on the trochanter. The 3 fatal cases with us which were not interfered with, took place after the Alma and Inkerman, and hence were probably wounded by conical balls.

All are agreed that, when practical, the separation of the limb should be accomplished at or through the **trochanter**, rather than at the **joint,** on account of the diminished risk; and this can be more often executed than would at first appear, as it not uncommonly happens that the fracture does not extend to the head of the bone, as it seemed at first sight to do; hence it might be judicious, in all doubtful **cases, to make** the incisions so low as to suit amputation at **the trochanter.** The steps necessary for exarticulation **can** easily be taken, if called for afterward, when the bone **is examined.** Such a proceeding would certainly **not be very** "brilliant," but it might save a life.

After the 1st April, 1855, **amputation in the upper third of the thigh** was performed 39 **times, with** a fatal result in 34 cases. Of the total number only one was a secondary operation, and it ended fatally. The ratio mortality per cent. was thus 86·8 for primary, and 100 for secondary. I have never myself seen any case recover in which the limb was amputated *beyond doubt* in the upper third, and I never met any one who had except in one instance, and that man was seen in England. I saw several upper-third amputations, so called, which were not really so. It is very easy to be deceived on this point. The French and Russians found these operations so hopeless that they almost abandoned them; and in fact, as was before remarked, the attempt to save such limbs, hopeless as it was, seemed more promising than amputation in the field.

Amputation in the middle third was performed during the period after the 1st April, 1855, 65 times, of which number 38 died; 56 of these cases and 31 deaths were primary operations, giving a ratio mortality per cent. of 55·3; 9 cases were operated on at a late period, and 7 died, or 77·7 per cent. **Amputation in the lower third** was performed during the same period 60 times, 46 being primary, **and 14** secondary operations: of the primary, 23, or 50 per cent., died; and of the secondary, 10, or 71·4 per cent. A very great many of the operations classed as "lower third" ought to have been entered as "middle third," as it very frequently happened that, from the operator adhering too closely to the maxim of Petit, to "cut as little of the muscle and as much of the bone as possible," an operation which was ostensibly in the lower was in reality in the middle third. This is a **matter** of which I have seen many illustrations; consequently I believe that at least one-third of the operations and the deaths **classed** as lower third should be transferred to the middle third column, and thus the relative frequency and fatality of the two operations would be better expressed.

Taking amputations in all parts of the thigh, then, we find the number of operations after the 1st of April, 1856, **was 164, of** which number 140 were primary, and 24 secondary operations. The total mortality was 105, or 64 per cent. Of the total deaths, the primary amputations yielded

87, or 62·2 per cent., and the secondary **18, or 75 per cent.**
It must always be borne in mind that these results only refer
to the period of the war when, as was before stated, **secondary**
operations were becoming very rare, **and the state of mat-**
ters in camp so improved **that** the total mortality after **am-**
putations was by no means what it had been **at an earlier**
period ; so that to say **that the average mortality after am-**
putation of the thigh, in **the** Crimea, was 64 **per cent., does**
not by any means express the whole truth. However, if we
take the later period only into consideration, then our results
may be thus contrasted with those obtained in other fields
of observation :—

TABLE SHOWING THE PERCENTAGE **OF DEATHS AFTER** AMPUTATION
(**PRIMARY AND SECONDARY**) **OF THE THIGH FOR GUNSHOT** WOUNDS
AND ACCIDENTS.

	Mortality per cent.
Crimea, British army from April **1st to end of** war..	64·0
Constantinople, French Dolma Batchi Hospital, Mounier...	82·6
Naval Brigade, Crimea	65·0
Indian campaigns	48·7
Waterloo	70·2*
Spain, **Alcock**	62·0
Sleswick-Holstein, Esmarch	60·15
Danish army, 1848-50, Djorup	56·7
Sédillot, "Campagne Constantine," 1837	87·5
Africa, Baudens	51·4
Polish campaign, Malgaigne	100·0
Mexican war	100·0
Hôtel-Dieu, 1830	81·8
Cases communicated to the Academy, **1848**	77·2

INJURY.

Phillips	71·8
Parisian Hospitals, Malgaigne	73·9

* I have in this computation taken for granted that one-third of
the cases "remaining" at the time the return given by Mr. Guthrie
was completed ended unfavorably, which appears a very moderate
allowance, when we find such a proportion as 51 out of 94 cases of
secondary amputation so entered, and 35 of the primary.

Glasgow, previous to 1848, Lawrie............................... 75·0
  "   M'Ghie .... 78·6*
  "   Steele .... 72·0
St. Thomas's Hospital, South................................... 85·7
Hussey.. 62·5
James, *all primary* .... 61·5
University College, Erichsen.................................... 60·8

The usual discrepancy which marks statistical tables is observable in the above enumeration. That between the results obtained in our army and those quoted from the French, and which were kindly furnished to me by M. Mounier, is easily understood when it is stated that of the total number of 46 amputations of the thigh which presented themselves in the hospital presided over by that distinguished surgeon, 25 were secondary operations, all of whom perished, while in our returns, and those of the Naval Brigade, there were very few consecutive amputations. Out of 21 primary amputations reported by M. Mounier, 8 recovered. The low mortality among the Indian cases is somewhat difficult to account for. In calculating them, I did not include any case except those the result of which I could find well authenticated. To distinguish between primary and secondary operations, in many of the cases recorded by the various authors referred to in the above table, was found impossible; but so far as this can be accomplished appears in the following table :—

---

* These numbers are derived from a further investigation of the Royal Infirmary records by Dr. M'Ghie and myself, and include the cases operated on for twelve years previous to 1853.

TABLE SHOWING THE MORTALITY AFTER PRIMARY AND SECONDARY (DISTINGUISHED) AMPUTATIONS OF THE THIGH FOR GUNSHOT WOUNDS

	Mortality per cent.	
	Primary.	Secondary.
Crimea, after April 1, 1855	62·0	75·0
Constantinople, Mounier	61·9	100·0
Legouest	......	100·0
Naval Brigade*	66·0	60·0
Indian campaigns	38·0	69·0
Spain, Alcock	64·7	60·0
Africa, Baudens	13·3	80·0
Cases communicated to the Academy, 1848, in which the distinction is drawn	57·0	81·2

TABLE SHOWING THE MORTALITY AFTER PRIMARY AND SECONDARY (DISTINGUISHED) AMPUTATIONS OF THE THIGH FOR INJURY.

	Mortality per cent.	
	Primary.	Secondary.
Malgaigne	75·0	60·0
Glasgow, Lawrie	91·6	66·0†
" Steele	65·6	83·6
" M'Ghie	61·2	96·6
St. Thomas's Hospital, South	100·0	50·0
University College, Erichsen	57·0	62·5
Hussey	83·0	......
James	61·5	......

* These numbers, as well as those given in the previous table, do not refer merely to the Naval Brigade as serving in the Crimea, but to the operations performed at the hospital on the Bosphorus.

† Dr. Simpson has completely confused Dr. Lawrie's **statistics**, having mixed up his primary and secondary **amputations** and those for disease. Thus, he gives 35 cases of primary amputation and 27 deaths, as occurring in **Dr. Lawrie's paper**, in which mention is made of only 12 cases and **11 deaths, out** of 35 cases and 27 deaths of primary and secondary cases *combined;* and under the head of secondary operations he has given those for disease.

If a calculation is made of the mortality succeeding amputation of the thigh from gunshot wounds *alone*, and the whole number of cases referred to in the above table included, then the average mortality per cent. of primary operations would appear to be 56·5, and of secondary 79·0; while, if the operations performed in civil hospitals for injury are alone calculated, then the average mortality of primary operations would appear as 69·6 per cent., and secondary 75·4—a result somewhat different from what is usually obtained.

Amputation through the knee-joint has been performed in our army 6 times primarily, 3 of which were fatal, and once secondarily, with a fatal result.* This very old operation has lately been creating some interest in the profession, and was often performed by the French surgeons in the Crimea. The opinion they were led to form of it may be supposed to be expressed by Baudens, when he says, (*Une Mission Médicale en Crimée:*) "It is a truth which the numerous facts observed in the Crimea permit us to affirm, that, whenever it is impossible to amputate the leg, the disarticulation of the knee should be preferred to amputation of the thigh. The former has more often succeeded than the latter." There are not, however, very many cases occurring in the field which are adapted for this operation, as it should be performed only when the injury is limited to the leg-bones and the femur remains intact; and when this takes place, it often happens that the soft parts are so much implicated as to deprive us of flaps. However, if the posterior flap is destroyed, we can take a long flap from the front, and *vice versa*. To 4 of the cases operated on in camp, with the details of which I am acquainted, the operation

---

* Of the 37 cases of this operation mentioned by Chelius as having been collected by Jäeger, 22 were favorable. Dr. Markoe, *New York Journal of Medicine*, January, 1856, gives the results of 18 operations performed since 1850 in America; 5 of these were fatal. If to these we add 6 cases which have been more recently published, with 1 death, we have a total of 61 cases and 21 deaths, or 34·4 per cent. mortality.

was not applicable, as the femur was more or less injured so as to call for the removal of part of it ; hence the operation, although termed amputation through the knee, was in reality low amputation of the thigh, such as that now employed in white swelling of the articulation.

As to the **mode of operation**, the French mostly adopted Baudens's method ; but in 5 cases operated on in the general hospital, that proceeding was departed from in so far as that the posterior flap was made from within outward in place of the reverse, as directed by that well-known surgeon. The anterior flap, too, was not made so long. Whatever method of operating be adopted, the great point which demands attention is to have the flap sufficiently broad to cover the expanded end of the femur, which there requires a large and broad covering. Of the 5 cases operated on in the general hospital, one died of phagedenic sloughing on the forty-third day ; another, a soldier of the 62d, died of enteritis on the sixty-seventh day, the stump being healed to a point ; a third sank from exhaustion on the ninth day after operation ; a fourth never fairly recovered from the shock ; while the fifth and last case recovered, under the charge of Dr. George Scott, who operated on him. This patient, a soldier in the Buffs, was struck on the right knee-joint by a ball, on the 8th of September. He thought himself very slightly injured, as the only thing he observed wrong with the joint was his inability to flex it, on account of "something catching in it." A small opening was found in the middle of the popliteal space, slightly external to the middle line, from which a good deal of blood flowed. This opening led into the cavity of the articulation, and spiculæ of bone were **felt** within. A part of the end of the femur was **removed, but** the patella left. A round ball had pierced the **external** condyle and lodged. The posterior flap eventually sloughed and exposed the end of the femur ; but the bone became subsequently covered over with granulations, and though the patient's progress toward recovery was much impeded

by the formation of an abscess among the muscles of the thigh, which required extensive incisions, he went to England in perfect health in January. His stump was strong and firm, and he had much power over its movements. The patella could be felt on the upper surface, to which position it had been gradually retracted. In several of the cases which I have seen in the French hospitals, where sloughing of the flaps had taken place as in this case, and exposed the extremity of the femur, the cartilages were alone thrown off, but not a scale of bone.

So far as I can judge, the practical advantages of this operation are equal in value to those theoretical ones which its advocates claim for it, and they would seem to recommend its more general adoption in any future campaign. First of all, the shock to the system is less, and we obtain a larger and firmer stump than when the femur is sawn through; the end of the bone on which the patient has to bear his weight is likewise more expanded and more rounded, and hence calculated to inspire greater confidence in the patient in the use of it, and less liable to cause ulceration by its pressure on its coverings.* A false leg can be more easily attached to such a stump, and more power is retained in progression from the muscles which remain undivided, than when the limb is amputated in the continuity. Few now participate in Liston's opinion of a long thigh stump, but, on the contrary, most surgeons try to keep their section as far as possible from the trunk. The non-interference with the medullary canal obviates many of the dangers of amputation, according

---

* The absorption of the condyles of the femur which may go on after this operation is illustrated by a case mentioned by M. Legouest, (*Amputation partielles du pied,*) in which a soldier had undergone amputation at the knee in 1800, in Italy, and "the enormous tuberosities had so diminished in volume that no trace of them could be recognized, but the member presented a cone terminated by a point." So completely had the part changed, that it was only after very careful examination they believed the man's story, that he had been amputated at the **joint.**

to Cruveilhier; while the extremity of the femur, which is largely supplied with blood-vessels, being retained, there is less fear of exfoliation than when the dense tissue of the bone has been opened by the saw. The position of the divided artery in the center of the flaps, and the few ligatures which are required, are further arguments in favor of this operation. There is little fear but that the flaps will adhere over the cartilaginous extremity of the bone—in fact, the cartilages soon disappear during the healing process. There is some appearance of force in the objection which some have advanced to the operation, that from the length of the stump no proper space is left for the play of an artificial joint; but if it be evident, as civil statistics at least prove, that the fatality attendant on this operation is less than that which follows amputation of the thigh, then any such objection loses all its weight.

If, then, cases were selected for the operation in which the femur remaining intact, and the leg-bones being destroyed, a sufficiency of flap could be got from the calf, **or** the front of the leg, and if the amputation was performed early, I firmly believe, with Malgaigne, that it is " Encore une de ces operations trop légèrement condamnées, et qui lorsqu'on a le choix mérite toute préférence sur l'amputation de la cuisse dans la continuité."

The **leg was amputated**, after April 1, 1855, 101 times, with death following in 36 cases, giving a mortality of 35·6 per cent.; 89 cases, and 28 deaths, were primary operations, and 12 cases, with 8 deaths, secondary—thus affording a ratio of mortality per cent. of 31·4 for the primary, and 66·6 for the secondary.

The rule generally followed in our army has, **I think**, been to preserve as much as possible of the **limb, but**, except in those cases in which the operation was performed just above the ankle-joint, the French **appeared** usually to amputate at the place of election. I saw no instance in which Larrey's operation through the head of the tibia was had re-

course to, but I am informed that it was several times successfully performed in the French ambulances.

The greatly improved mechanical contrivances of late years
have much changed the bearing of the question with regard
to long leg stumps. The facility and moderate cost with
which artificial limbs can now be fitted to any part of the
limb, from the knee to the foot, has obviated many of the
reasons which formerly induced surgeons to prefer the high
operation. Larrey's, through the head of the tibia, is a
most valuable one when the destruction has extended high
up the leg, as it enables us to retain the use of the knee-
joint, as well as diminish the risk to life. That at "the place
of election" will, of course, continue to be employed in cases
of injury above the middle of the leg; but when the nature
of the accident permits of it, the part of the leg which appears to combine most of the advantages sought in leg
stumps by both the surgeon and the mechanician, is undoubtedly that in the center of the middle third.   The
length of the lever thus obtained, the diminished bulk of the
part, and consequently of the truncated section, the means
of covering the bones, and the room it affords for attaching
a limb, are all in favor of this locality. Many most admirable stumps were made in this part of the limb during
the war. In operations for accident, as in gunshot wounds,
we can, of course, operate lower in the leg than we can
when the operation is undertaken for disease, from the absence of the thickened state of the bone, and the changed
and bound-down tissues which are so common in cases
operated on in civil hospitals.

As to the operation just above the ankle, which has of
late years caused so much discussion on the continent, we
had, so far as I know, no experience in our army; but the
French had a good number, which, so far as the condition
of the stumps went, were not by any means promising. This
operation, although revived by the improved method of
procedure introduced into practice by M. Lenoir, is yet of

sufficiently old date. It is mentioned by Dionis in his "Cours d' Operations," and was practiced by Bromfield in 1740, and afterward by White, Alanson, and Bell, in England. In France, Blandin often performed it in recent times, but was induced to abandon it, like many others, from the bad results his method of operation yielded. By M. Lenoir's modification,* and M. Martin's artificial limb, the operation promises again to come into favor. This **operation** appears to me to have a special bearing on military practice. Its value will be best judged of by considering—1st, its safety; and, 2d, the usefulness of the resulting stump. As to the first point there can be no question as to its advantage over any other amputation in the leg. The greatly diminished bulk of the soft and hard parts at the place of section, the smaller amount of shock such severance will occasion, and the more rapid closing of the wound, are all incontestable. Its fatality in the cases operated on in France has been only as one-sixth or one-seventh, while the mortality of amputation at the place of election is more than one-half, (55 in 100 according to Malgaigne.) In some hospitals, as in the Beaujon, the mortality has been even less in the *sus-malléolaire* operation than that mentioned above: thus M. Huguier only lost 1 out of 14 cases. So, then, as far as the mortality goes, there can be no division of opinion, as there is about the second point, viz., the state of the stump afterward. The difficulty of retaining enough of covering for the bones, the fear of such retraction as will occasion a conicity of the stump, the tenderness of the cicatrix and its inability to stand pressure, the chance of fusiform collections of pus forming among the tendons, **of** caries or necrosis of the bones following,—all these are among the objections which have been advanced **to the** operation. If we, **however,** carefully examine these by the

* See Arch. Gen. de Med., July, 1840, and Mémoire by Arnal and Martin, Paris, 1842.

light of the large number of observations which can now be brought to bear on the subject, we find that the only objections which are of any weight are the scanty covering of soft parts, the tenderness of the cicatrix, and the risk of necrosis. Purulent collections can be easily avoided by careful dressing; and the presence of the other evils, and, in fact, the want of flap also, must be referred to the manner in which the operation has been performed. I have examined a considerable number of those amputated in Paris, and am bound to say that, while in some cases the evils spoken of existed, in the greater number of instances good and firm stumps were formed. This was especially the case in several which I saw in M. Lenoir's service, in the Neckar. Some of the cases which had been operated on in the Crimea were certainly very bad. At the Society of Surgery I saw an Arab, shown by Baron Larrey, both of whose limbs had been removed above the malleoli, in the East. They were both secondary operations, and seemed to have healed well at first; but the cicatrix afterward ulcerated, and at the period he was shown to the society (nearly two years after operation) he could not use his stumps in any way, from their being in an unhealthy condition. In another case, shown to the same society on a subsequent occasion, the operation had been performed in 1848, and the man had been an inmate of hospitals on several occasions during the interval, on account of ulceration, abscesses, and necrosis in his stump. The bones were much thickened, and evidently diseased at the time I saw him. A letter from M. Hutin of the Invalides, which was at the same time read, stated the results of the operation as they had come under his observation, and certainly his evidence was not favorable; however, the want of a properly constructed artificial limb for the patients detracted much from the value of his remarks.*

---

* Larrey lost several of his low amputations by tetanus, which must have been a mere coincidence. Ballingall tells that " of 34 sol-

If the limb cannot be fitted with a false foot, but made to
rest on the knee, scarcely anything will make amends for
the long and cumbrous stump. Since 1845 M. Hutin had
had 5 cases especially under his notice: one could walk,
but with difficulty, and would willingly part with his foot;
one had been several times in hospital from the state of his
stump, and three had to undergo subsequent amputation.
Now all this is sufficiently distressing and discouraging, but
in military practice I question whether it is conclusive. The
limited mortality yet presents itself to us as a great fact,
which arrests our attention. If when men die so fast after
the ordinary amputation of the leg, as they did during the
early part of the war in the Crimea, it becomes a grave consideration whether, with all its subsequent drawbacks, we
should not adopt this process when practicable. If our
choice lay between two operations of equal gravity, then
unquestionably we are bound to select that which will provide the most useful stump; but when the chances of death
are beyond all comparison greater in the one case than in
the other — when, independently of those dangers which
attach to the operation itself, the marked presence of a
hospital epidemic makes it desirable to expose as small
and as rapidly-healing a surface as possible, then I think it
may be conceded that the *sus-malléolaire* operation has
many claims upon us. Life must be our chief concern,
convenience a subordinate consideration. The complaints

---

diers admitted into the Invalides, after the Russian campaign, with
their legs amputated immediately above the ankle, 22 had such bad
stumps as to induce them to submit to a second amputation below
the knee." I heard Baron H. Larrey inform the Société de Chirurgie
that the Russian surgeons employed in the same campaign had informed him that hospital gangrene being very rife in their army, they
adopted the low operation, so as to leave room for a subsequent one
if the stump went wrong! In the text I have alluded to the use of
the operation for injury only. It is not thought applicable to cases
of malignant disease of the foot, from its near neighborhood to the
affected part.

of patients about the inconvenience of their stumps must be considered as affording little evidence in the matter, as the fact that they survive to murmur is often due to the very operation against which they complain.

If the heel remains, then this operation could not be thought of; but it is in those cases, sufficiently frequent in their occurrence, in which the whole foot has been carried away by round shot, or such like accident, and in which the choice of operation lies only between the amputation above the malleoli or higher up, that the merits of this method can be weighed. The careful study of those cases in which caries or necrosis has appeared in the bones of the stump, after the *sus-malléolaire* amputation, will be found to have been submitted to operation for disease, and not for injury, and it will generally be found, besides, that a faulty apparatus has been used afterward. Everything depends on the careful adaptation of the false foot, and, so far, this is of itself an objection to the operation being performed on the poor; but the view alone I wish to take of it at present is with reference to military practice, and there it seems to promise many advantages at times when there prevails a high mortality after operations.

Amputation at the ankle-joint was performed 12 times in the Crimea during the period embraced by the returns, and death followed in 2 cases. Of the total number of cases 3 were secondary operations, and these were all successful. Syme's operation was as useful and as successful in its results as usual. Pirogoff's modification of Syme's method was, I understand, several times tried at Scutari. I saw none of these cases, and am ignorant of the results. In England it appears to have been recently followed by good effects in 6 out of 9 cases in which it was performed. Langenbeck is said to approve of its results in a good many cases in which he has tried it; but the history of the 3 cases first reported by M. Pirogoff himself, and those more recently put on record by Michäelis, of Milan, and various German

surgeons, does not hold out much encouragement to repeat
the operation, not only from the long period necessary to a
cure, but also from the unsatisfactory nature of the resulting
member.   It was reported in the East that this operation
had been frequently performed by Pirogoff himself in Sebastopol, but that he had found the calcaneum act as a foreign
body in the stump, and was hence disposed to abandon it.
Roux of Toulon's operation was performed once in the
general hospital in camp, with most excellent results.   The
chief objection to this operation arises from the vessels and
nerves being drawn under the bone ; however, it certainly
enables us to form a stump little inferior to Syme's, when
the half of the heel has been destroyed.   Baudens is said
to recommend the flap to be taken from the anterior surface
of the joint, or even from its external surface, **if** it can be
got no other where, rather than go above the ankle.   Chopart's operation was performed primarily 7 times, one case
ending unfavorably, while Lisfranc's was successful **in** the 4
cases in which it was tried.   The step now always followed
by Mr. Ferguson, of removing the projection **of the** astragalus in performing Chopart's operation, is an undoubted
improvement.

The **upper extremity** has been removed at the **shoulder-
joint**, between the 1st of April, 1855, and the end of the
war, 39 times, with a fatal issue 13 times, or 33·3 per cent.
Of these operations **33** were primary, and 9 deaths followed,
giving thus a mortality of 27·2 per cent.; while of 6 secondary
operations 4 died, or 66·6 per cent.   During the previous
period of the war, at least 21 other cases of amputation at
this joint were performed, beyond the 39 mentioned above,
and of that number 6 died, thus presenting **a** total **of** 60
cases and 19 deaths, or **a ratio** of mortality of **31·6** per
cent. overhead.*

* The mortality following this operation is shown in the following
table.   Larrey is said to have had upwards of 90 recoveries from

It is impossible fairly to contrast the results of amputation at the shoulder and that in the shaft of the humerus, as, in military practice particularly, it very much oftener happens that the trunk has suffered severely in those injuries which necessitate exarticulation than those in which amputation of the upper arm alone is required. Not a few illustrations of this occurred in the Crimea. Thus, in at least two of the cases returned as shoulder-joint amputations, besides the in-

---

about 100 cases on which he operated, and this success he attributes to his method of operating.

Amputation at the shoulder.	Primary.	No. of deaths.	Secondary.	No. of deaths.	Ratio mortality per cent.
Crimea, after April 1, 1855	83	9	6	4	33·3
French, Dolma Batchi Hospital..........................	3	1	6	3	44·4
India............................	4	1	....	....	25·0
Sédillot, Constantine.........	2	2	....	....	100·0
Larrey, (fils,) Antwerp......	5	....	3	2	25·0
Alcock, Spain.................	9	1	1	1	20·0
Guthrie, Spain.................	19	1	19	15	42·1
" Waterloo............	6	1	12	6	38·8
Roux, 1848.....	3	1	....	....	33·3
Larrey, Gross-caillou, 1830	....	....	2	1	50·0
INJURY.					
New York Hospital..........	7	8	7	6	64·2
Glasgow, Lawrie..............	3	2	1	....	50·0
" M'Ghie, 1842–53	17	6	5	8	40·9

Malgaigne also reports 7 cases, all ending fatally, but does not indicate whether they were primary or secondary operations. If, then, we sum up all the cases in this table, we find that the ratio of mortality is 20·2 for primary operations performed in the field, and 65 for secondary; while both operations in the field yield a mortality of 36·8 overhead. If, on the other hand, we calculate the operations for injury alone, then those performed early give a ratio of mortality of 40·7 per cent, and the secondary 69 per cent.; while overhead operations for injury give an average mortality of 50 per cent., and primary and secondary operations, for both gunshot wounds and civil accidents combined, yield a total average mortality of 39·8.

jury to the arm, the scapula was carried away or destroyed, and the muscles of the chest torn.

In no operation is the advantage of **primary** over **secondary** amputation so evident as in that at the shoulder-joint, early operation at this part being an exceedingly successful undertaking, while late interference generally affords a considerable mortality. Thus, if we take Guthrie's experience **in Spain,** and Dr. Thomson's observation after Waterloo alone, this point is well illustrated ; of 19 cases of secondary amputation mentioned by Guthrie as having been performed between June and December, 1813, 15 died, while of **an** equal number who were operated on in the field only 1 died. Dr. Thomson again says: "In Belgium almost all of those **recovered who** had undergone primary amputation at the shoulder-joint, while fully one-half died of those on whom it became necessary to operate at a late period." The same point is illustrated to some extent by our Crimean results, less than a third of the primary and two-thirds of **the** secondary perishing.

Deputy-Inspector Gordon **had one case of** recovery, in which both the arm and the greater part of the scapula were removed. Mr. Howard, of the 20th Regiment, successfully removed the right arm of one man and the left of another, in close succession, at the joint, for injury occasioned by **the** same cannon-ball, which had struck between them.*

* The following is **a most** instructive case, as showing how the operation of amputation at the shoulder may be recovered from under the most unpromising circumstances. It occurred in the 29th Regiment, serving in India, and under the care of Deputy-Inspector Taylor. Sergeant Ritchie was struck by a cannon-ball on the **upper** part of his left arm, by which the bone, including the head **and upper** third of the humerus, **was** smashed. Both folds **of the axilla** were carried away, and the **artery was divided.** The arm **was** only kept attached by a portion **of the deltoid and the skin** covering it, and of these the flaps were **made.** This man lay exposed on the field for three days; yet **he recovered** completely. "His case is **peculiar in**

Amputation of the upper arm was performed in the Crimea, from April 1st to the end of the war, 102 times, followed by death in 25 cases, the mortality per cent. being thus 24·5. Of the total number, 96, and 22 deaths, were primary operations. The ratio of the mortality was thus 22·9 for the primary, and 50·0 for the secondary operations.

The forearm was amputated during the same period 52 times *primarily*, and the hand at wrist once, with only one death; while of 7 *secondary* operations, in the same parts, 2 died.

These returns do not speak of a considerable number of secondary amputations of the arm, which were performed early in the war, and the success of which was certainly such as to warrant us in trying to save, in the first instance, most cases of gunshot wounds of the arm. It is almost impossible to say what wound of the arm by a ball will not recover; so that it is a well-recognized rule to wait, in all but desperate cases, and only amputate, if unavoidable, at **a subsequent** period. In military practice, secondary amputations are only justifiable when performed on the upper extremity.

The **mode of managing stumps** in the East was that usually followed at home for the promotion of adhesion by the first intention. The edges of the flaps were usually united by **suture.** The observation of this method in the Crimea did not certainly appear to be satisfactory. To wait, as Liston so strongly advocates, till all oozing has ceased from **the cut surface,** is unquestionably a most useful precaution, and one of great moment to their successful and early union. The **irritation which the** stitching of the edges occasions, the want **of sufficient** room for subsequent swelling, the confinement **of pus which is thereby favored, all** appear reasons

---

two respects: 1st, no ligature was needed; and, 2d, at least two-thirds of the face of the stump was the surface left by the passage of the cannon-ball, and yet it healed very kindly." Dr. Taylor informs me that he recently saw this man in good health. He is on the staff in Belfast.

against sutures. Stripes of wet lint applied like adhesive plaster always appeared preferable. I never saw one case among our most numerous amputations in which primary adhesion took place throughout the whole surface of the flaps. They united readily enough along their edges; but the result of this was that a large bag of pus was formed within the end of the stump, which continued as a depot for absorption into the system, by steeping the end of the sawn bone and the vessels in its matter, and it burrowed far and wide in the intermuscular spaces and along the bone, and ended not unfrequently in causing considerable necrosis of the end of the divided shaft. Unquestionably it may be said that such collections should have been recognized and prevented; but yet it seems to me that when ample proof is afforded, as it was early in the East, that primary adhesion was the rare exception, and not the rule, and when the patients were so peculiarly liable to purulent absorption as they were with us, it would have been better practice not to have attempted primary union, but to have adopted such treatment as best favored the freest discharge of the matter so soon as it was formed.

The method of dressing with compresses, recommended by Mr. Luke, was most useful, in several cases in which I tried it, in preventing the accumulations referred to. The contrast afforded by the heavy dressings for stumps, employed by the French, and our water dressing, was very marked, and may have contributed something to the result which obtained in the less prevalence of purulent absorption in our hospitals than with them. Bad as it was with us, it never became the terrible epidemic it was in the French hospitals. We had no means of trying the method of treating stumps in water, recommended by Langenbeck. The ease with which the purulent secretion can be got quit of by position, in amputations of the arm and leg, contributes, I have no doubt, not a little to the decreased mortality attending these operations, as compared to amputations of the thigh. The Russian surgeons, I am told, when operating by the circular method, which they generally adopt, split the posterior flap, and keep this part open in order to drain off the pus. Such a step would meet with little favor in this country, but it presents many advantages when purulent absorption is so common as it was in the East. M. Sédillot, of Strasburg, I believe proposes a similar modification for general use.

Primary adhesion is, of course, most desirable when hospital gangrene prevails, but it is just at such a time that this result is most difficult to obtain.

Cases of secondary amputation of the thigh for injury of the knee were always those in which attempts at primary union did worst. The long fusiform collections of matter, which are so apt to exist in these cases previous to operation, extended, and did every possible harm. Careful bandaging from above downward to the base of the flap seemed to be highly useful in these cases.*

Pus poisoning was unquestionably the chief source of our mortality in the East, after amputation, especially after secondary operations. The resemblance between its early features and those of ague was perhaps more marked among our patients than it even usually is. This seemed especially the case among men who had served during the early part of the war—of this, however, I am not certain. We had many most beautiful examples, post mortem, of veins leading

---

* Sir Charles Bell strongly advocated the bandaging of stumps, "to compress the veins and cellular membrane, so that the adhesive inflammation in the mouths of the veins may prevent the inflammatory action on the face of the stump from being communicated to the great vessels. The great vein," he says, "being properly compressed, adheres, and otherwise it lies loose and open, and the inflammation of the general surface will be communicated to it." It is not for this reason, but to oppose the burrowing of matter, and to prevent muscular contractions and the protrusion of the bone, that it is now adopted.

from the stump remaining round, patulous, and filled with pus, and sometimes reddened in their interior. It was not uncommon to trace the pus-filled vein from the thigh to the vena cava.

It is a question on which it **is** difficult to decide whether or not, when pus absorption is so common as it was with us, it would not be justifiable practice to ligature the chief vein at the time of operation. The views of Mr. Travers and others would certainly seem to oppose the adoption of such a step, but we have, on the other hand, the evident absorp-**tion** of pus **into** the system by this channel; and, besides, numerous cases are on record in which the ligature of the vein has not only not **been** followed by evil results, but has absolutely been the apparent cause of preventing inflamma-tion **and pus absorption.*** The non-appearance of symp-toms of purulent poisoning till after the separation of the threads makes it generally difficult to say which set of ves-sels—those ligatured or those not ligatured—have been the carriers of the pus. In the case referred to in **the note** death took place rapidly, before the ligatures were detached. Hennen expresses himself thus on the danger of tying veins: "When the great veins bleed I have never hesitated about tying them also, and it is most particularly necessary in de-bilitated subjects." Chevalier, too, says: "I know from ex-perience that the **principal vein of** a limb **may** be included in the same ligature as the artery without any disadvantage ensuing." Every hospital surgeon has seen instances of the same thing. I most firmly believe in Stromeyer's views on absorption by the veins of the bone, from observations which have been presented to me.

Independently of all fortuitous circumstances, there can

---

* This is particularly well illustrated in a case related by Mr. Johnston, of St. George's Hospital, **in the** journals of 1857. In that case those vessels which had been **tied were free both** of inflamma-tion and pus, while those not included in ligatures were full of pus, and "**much inflamed.**"

be little doubt but that some constitutions oppose themselves more to pus poisoning than others. This, although a most unsatisfactory mode of explanation, yet seems the only way of answering the difficulty which is presented to us in the much greater susceptibility of some to purulent absorption than others. Most die rapidly, while others, not apparently so well fitted to withstand the assaults of such an invader, though placed in precisely the same circumstances, only yield inch by inch, and others again escape altogether.

The presence of typhus fever in a hospital has been supposed to favor the development of pyæmia, and, although it cannot be denied but that the diseases often coexist, yet it seems more probable that they both proceed from a like source—a lowered vital energy in the patients, or vitiated hygienic arrangements.

The secondary deposits were with us, as usual, generally found in the lungs. Beck states, as the results of his observation in Holstein, that such was the seat of the deposition in seven cases out of ten in which patients died of pyæmia. This is not, I believe, an exaggerated average. Some of the French surgeons employed at Constantinople made the remark that they seldom found the pus collected in depots, as they had been accustomed to see it in Africa; but that it commonly was disseminated through the organs, muscles, and bones.

The visceral congestions which so often follow amputation were more than commonly fatal in their results in the Crimea, from the presence in most cases of the seeds of disease in the lungs, kidneys, and intestines. Phthisis and acute dysenteric attacks were not unfrequently the immediate causes of death, and in at least two cases the symptoms of Bright's disease of the kidney were most rapidly developed after thigh amputations.*

---

* For an outline of the statistics of the French army, see Appendix II.

APPENDIX.

30*

APPENDIX.

THE following summary of the geology of the allied position is from a Report published by Dr. Sutherland, the Government Commissioner to the seat of war :—

"The geological series, from above downward, includes the following formations : 1. The newer tertiary, or steppe limestone. 2. Volcanic cinders and ashes. 3. The older tertiary. 4. Nummulitic limestone. 5. White chalk and green sand. 6. Neocomien. 7. Jurassic limestone. 8. Conglomerates. 9. Schists. 10. Erupted volcanic rocks.

"1. *The newer tertiary limestone* forms the superficial stratification of the plateau before Sebastopol, and also the higher levels of the country to the north and northeast of Sebastopol harbor. The siege-works were principally excavated in it. This limestone affords good rubble building stone, and also an inferior road material.

"2. Immediately under the upper tertiary beds at San Georgeo is *a bed of volcanic ashes* containing shells, which can be traced from the great ravine of San Georgeo along the sea-coast to Cape Chersonese, and thence round the inlets of Sebastopol harbor to Karabelnaia.

"3. *Older tertiary beds* underlie the volcanic ashes in the cliffs of San Georgeo. They come to the surface at Karabelnaia, and form the Heights of Inkerman, as also the hills bounding the north side of Sebastopol harbor.

"4. *The nummulitic limestone* forms the hill-slopes and cliffs of Inkerman, in the ravines of which it has been extensively quarried for building stone. The hill-slopes above the quarries are covered with loose nummulites. The formation again appears in the hills at the head of Sebastopol harbor, extending from thence to the northeast of the line of Mackenzie's Heights.

"5. *The white chalk* begins, on the west, at the ruins of Inkerman, where it is mixed with green particles and upper green sand fossils. It forms the line of cliffs and talus of Mackenzie's Heights: also the bed of the lower valley of the Tchernaya, and occupies the area between the slopes of Mackenzie's Heights and the ridge which separates that valley from the basin of Balaklava. It extends eastward along the base of the heights, and fills up the space between them and the jurassic limestone group east of Tchorgoun, rising into round-backed lofty hills. It forms also the line of hills south of the Tchernaya, known as 'Fedoukine Heights.'

"6. *Neocomien beds* appear under the chalk near Tchorgoun, and extend along the western side of Schula valley toward Aitodar.

"7. *Jurassic limestone* appears on the west, in the great cliff at the ravine of San Georgeo. It forms the sea-coast cliffs and mountain chains to the eastward, and also the mountain groups between the valley of Tchorgoun and the Baidar and Varnoutka basins. The rock is much altered, dislocated, stratified, hard, and compact, often fissured, and the fissures filled with indurated red clay. Not unfrequently it caps the conglomerate.

"8. *Conglomerates* of different degrees of fineness occur from the ravine of San Georgeo to Baidar valley. Fine-grained beds of conglomerate, apparently altered by heat, underlie the jurassic cliff at San Georgeo. Immediately to the northeast of the cliff the formation reappears, and forms part of a chain of hills closing the upper end of the valley of Karani. The hill on the south side of the entrance to the valley above the bazaar at Kadikoi also consists of the same formation. Marine Heights and the hills to the east are wholly or partially formed of conglomerates, as are also the southern and western slopes of the Varnoutka basin. Part of the mass of Cape Aia consists of the same rock.

"9. *Schists*, apparently belonging to the Lias, underlie the conglomerate beds in the ravine of San Georgeo. They reappear on the south and eastern sides of the basin of Balaklava, under the

Col. and in the ridge separating Balaklava basin from the valley of the Tchernaya. They are found in large masses in the valleys to the east of Kamara, from whence they extend southward to the sea-shore. They occur in the basins of Varnoutka and Baidar, and in the undercliff below Laspi.

"10. *Erupted volcanic rocks* form the vast picturesque masses of Cape San Georgeo. They underlie the upper and lower tertiaries there, and they protrude themselves at intervals among the jurassic limestones and schists along the south coast of the Crimea to the eastward."

APPENDIX B.

"Observations were kept irregularly by various persons in Balaklava, but there was no regular series except those kept at the Castle hospital by Drs. Jephson and Matthew. The instruments made use of were an aneroid barometer, a maximum and minimum thermometer, a wet and dry bulb thermometer, by Negretti and Zambra, a sun thermometer and an air thermometer. The instruments were placed on the north side of one of the huts, about 320 feet above the sea, and overhanging it. From this circumstance, and from partial observations elsewhere, it is probable that the Castle hospital observations represent a sea climate rather than a land climate; that the mean temperature in the close land locked harbor of Balaklava, with its overhanging mountain slopes reflecting the sun's rays, was higher than at the Castle hospital, at least during summer; and that the extremes of heat and cold, as well as of dryness, were greater on the plateau before Sebastopol.

"The highest observed sun temperature was on the 14th August, 1855, on which day the sun thermometer indicated 125° F. The highest observed shade temperature was 99° F., on the 23d July; and the lowest observed temperature was 2·5° F., on the 19th December, 1855.

"On comparing the climate of the allied occupation with that of the metropolis for a series of years, we find that in April, 1855, the excess of mean temperature at Balaklava over Greenwich was 3·8° F.; in May, 9·5° F.; in June, 11·9° F.; in July, 11·3° F.; and in August the excess was 11 9° F. In September, 1·8°

F.; in October, 9·4° F.; in November, 4·6° F.; in December, the Crimean temperature wes 7·1° F. under the London mean of the month. It was 1·7° F. above the London mean in January, 1856. In February it was 2·6° below the London mean, and 9·2° F. below the same mean in March. In April the Crimean temperature showed an excess of 1·4° F., and in May, of 7·5° F. above the London mean.

"The daily mean range of the month was in excess of that of Greenwich. In April, 1855, the excess was + 4·4°; in May, + 4·1°; in June, + 4·1°; and in July, + 5·4°. In August it was + 4·5°; in September, + 1·8°; in October, + 5·1°; in November, + 3°; in December, + 2·3°. In January, 1856, it was + 1·2°; in February, + 1·9°; in March, + 1·5°; in April, + 1·2°; and in May, + 1·1°.

"The following table gives the monthly means and ranges, from April 1, 1855, to May 31, 1856, as deduced from the observations kept at the Castle hospital, Balaklava:—

Month.	Barom. mean.	Barom. range.	Mean temp.	Mean daily range.	Mean max.	Mean min.	Mean dry.	Mean wet.	Mean sun temp.	Days of sunshine.	Rain.
	Inches.	Inches.	Deg.	Deg.	Deg.	Deg.	Deg.	Deg.	Deg.		Inches.
1855.											
April............	29·433	·962	50·3	21·8	64·1	40·7	57·0	50·0	68·1	22	2·846
May.............	·544	·748	62·9	23·2	74·8	51·3	64·6*	58·5*	81·0	29	5·308
June†............	·624	·480	71·2	23·9	83·6	59·8	......	......	96·4	29	3·825
July.............	·543‡	·474‡	73·1‡	22·9§	84·7‡	62·2‡	76·6	67·9	93·32	29	4·008
August.........	·574	·385	73·0	22·3	84·5	61·5	70·6	68·9	107·5	28	2·776
September.....	·634	·535	58·6	19·1	68·0	48·7	61·2	51·9	85·9	23	No data.
October........	·610	·540	53·1	18·7	70·5	50·0	61·0	55·6	81·4	27	·118
November.....	·651	·870	48·9	13·7	54·9	41·0	49·8	45·7	82·4	16	2·057
December.....	·568	·930	33·3	11·3	39·3	28·0	35·1	33·0	53·2	18	2·490
1856.											
January........	·499	·760	40·0	9·4	46·0	35·0	40·8	38·0	59·0	15	2·499
February.......	·536	·715	35·2	12·4	42·5	30·0	38·0	35·7	58·8	15	2·438
March..........	·500	·870	32·6	15·4	40·5	25·0	35·9	33·2	65·8	22	2·012
April...........	·481	·465	47·9	18·1	56·4	38·9	50·5	44·27	84·9	26	1·203
May.............	29·408	·605	60·9	20·2	71·0	50·7	62·9	56·0	85·4	26	1·329

* 12 days.   † 29 days.   ‡ 28 days.   § 29 days.   ‖ 21 days.

"So far, then, as can be ascertained by the observations, the Crimean climate, during the period of the allied occupation, may be characterized as one of extremes,—intense summer heat and sun radiation, and severe winter cold. The observed difference of air temperature in July and December was 93·5° F.; and the difference between the highest sun temperature and the lowest air temperature was 122·5° F. The daily variations were also at times excessive. During the hot season, the daily maximum shade temperature ranged from 72° to 99° F., while the minimum ranged from 44° to 72° F. The sun temperature, to which the troops were exposed day after day during the same season, varied from 110° to 125° F. The passage from the sunshine to the shade was attended by a fall of temperature of from 32° to 44° F. A sun temperature of 120° F. was followed by a fall of from 50° to 60° F., at the minimum period of the same night.

"The barometric means, so far as could be ascertained by the aneroid barometer, were steady, and the range under one inch.

"The rainfall in November, 1855, was 3·167 inches, all of which, except less than half an inch, fell on the last six days of the month. There was a little snow on the 27th Between the 1st and 12th December there fell 2·300 inches of rain. There was a little snow on the 13th, and six inches of snow on the 17th and 18th together. During January, 1856, there fell, on thirteen days, 2·499 inches, and about seven inches of snow fell on the 4th, 5th, and 17th of the month. Snow fell on eight days in February, the heaviest fall being on the 1st of the month, and equivalent to 1·294 inches of rain. There was hardly any rain during the month, but the total fall of snow was equivalent to 2·438 inches. The greatest cold of the year was on the 19th December, 1855, when the minimum temperature was as low as 2·5° F. The maximum of the same day was 9° F. The mean temperature in November was 48·9°; in December, 33·3°; and in January, 1856, 40° F."— *Sanitary Commission to the East.*

APPENDIX C.

Table by Dr. Christison, showing the weight and nutritive value of each article issued to the British and to the Hessian soldier as a daily ration :—

	Ounces of nutritive principle.	Whereof there is	
		Carbonif-erous.	Nitro-genous.
British sailor, daily nutriment, exclusive of beer.................. }	28·5	20·90	7·54
Hessian sailor, daily nutriment......	32 96	26·59	6·37

British soldier in the Crimea, receiving daily—
    1 lb. salt meat.................. }
    1 lb. biscuit..................... }
    2 oz. sugar...................... }
    Coffee, not used; rice, uncertain; beer, none.
	23·52	16 6	6 92

APPENDIX D.

"Return showing the duty performed by the army in January, 1855:—

Rank and file.	Brigade of Guards.	2d Division.	3d Division.	4th Division.	Light Division.	Total.
Effective and present under arms..	948	2469	2668	2332	2770	11,367
Detailed for duty of various kinds daily..............	403	827	1170	1431	1490	3321

"The results for December and February were much the same as in January."—*Col. Tulloch*, p. 176.

APPENDIX E.

"The routine **of** duty in particular regiments is thus described by various officers:—

"Lord West, commanding **the 21st Regiment, states** that:—

"'Those for the day covering party **are** roused out of their tents at 4 o'clock in the morning, have about a mile and a half **to** march down, through snow and mud, **and** get back to their camp about 7 o'clock in **the** evening, being thus exposed, in open trenches for 15 hours, to such inclement weather as now prevails. Most of them will go on the following evening at 5 o'clock, and remain out all night till 6 o'clock the following morning; this routine has been kept up incessantly for the last six weeks.'

" Lieutenant-Colonel Maxwell, commanding the 46th Regiment, a corps which was nearly annihilated by sickness in the months of November and December, states that the number of hours his men were in the trenches in every 24, was 12 in the first of these months, and 10½ in the second; and it was stated by the surgeon, and verified by the lieutenant-colonel, that at one time the men were in the trenches for six successive nights, and had only one night in bed in the course of a week, but that afterward the duty was better regulated.

" 'The duties in the Light Division are thus described by Deputy Inspector-General Alexander, in a letter dated 10th December, 1854:—

" ' In the 7th Fusiliers, men were in the trenches 24 hours without relief, up to or about the 17th November; on the 14th two companies were kept on picket for 36 hours, when, of course, no cooking took place.

" ' In the 19th Regiment, taking the total number of hours for November, viz.. 720—304 have been passed by the men either on duty in the trenches, or on picket, which is 10 hours daily for each man, the remaining fourteen being passed in bringing water, seeking for fuel, cooking, and other duties, etc. In the 23d Fusiliers, the average return gives to each man one night in camp and one on duty; many men, however, had to go on duty with their companies two or three nights running, doing 24 hours' duty to 12 in camp.

" ' In the 34th Regiment the men, on an average, were something less than one night in their tents, with water and fuel fatigues when off duty; they are, in consequence, weak and wasted from the incessant and severe duty.

" ' In the 77th Regiment the men were either in the trenches or outlying picket every second night; on the intervening days, guards, besides water and fuel fatigues, etc.

" ' In the 88th Regiment no man has ever more than one night in his tent, has 12 hours in the trenches, and 24 hours on picket, and then has to look after wood for cooking, water, etc. etc.'

"A return and letter from Captain Forman, commanding the

right wing of the 2d battalion of the Rifle Brigade, also shows that in November that wing was on duty 17 times, namely, 9 in the trenches, and 8 in picket; and that the average daily duty performed by each man was about 10½ hours, in addition to two hours spent in going to and from the trenches, besides the fatigue of procuring wood and water, and other regimental duties.

"In December the amount of duty in that corps is described as being rather less, viz., only about 9 hours in the trenches or pickets, exclusive of other duties.

"These few individual instances will be sufficient to show how the system worked, and there appears no reason to suppose that (except, perhaps, in the 46th Regiment) they differed from the ordinary routine of duty in other corps during this period."— TULLOCH, *Crimean Commission*, pp. 177–78.

APPENDIX F.

TABLE *giving the weight of the different balls used by the various belligerents during the past war. With the exception of our own the weights noted can only be looked on as close approximations, from the specimens being too few to enable the weights to be given with perfect exactness. All the specimens examined were new balls which had not been used.*

Description of ball.	Average weight.			
ENGLISH.	Ounces.	Drachms	Scruples	Grains.
1. Large conical ball with a cup. Ball of 1851. Used now, I believe, only by the marines...........	i	iij	i	xiij
2. Do. without the cup..................	i	iij	....	x
3. Ball of 1853. Longer cone than the last. The Enfield rifle-ball Hollow in the base. Coming into universal use in the infantry.......	i	....	ij	i
4. Conical ball with an iron cup. Same size as the last, and formerly used for the same rifle	i	....	ij	v
5. Old round musket-ball..............	i	....	....	x
6. Lancaster ball. Small elongated cone. Used by the sappers.........	....	vij	....	ix
7. Old rifle-ball, with belt, 1842, not now in use.................	i	i	....	xiv

Description of ball.	Average weight.			
	Ounces.	Drachms	Scruples	Grains.
FRENCH.				
1. "Balle à culot." Large conical ball, with three rings and a deep cut. Used by the Zouaves	i	vj	....	vj
2. Minié ball. Used by the grenadiers and voltigeurs of the guard..	i	ij	ij	iv
3. "Balle carabine a tigne." Used by the chasseurs de Vincennes.....	i	v	....	xvj
4. Small conical ball of the artillery of the guard. Two deep rings.....	....	ij	ij	xvj
5. Round infantry ball.....................	....	vj	....	viij
6. Variety of last, being a half sphere with a hollow base—not so much used.................	i	....	....	xvj
SARDINIAN.				
1. Carabine of the Bersagliere. Conical ball, with solid base, and coming rapidly to a very sharp point..	....	jxss	....	....
2. A short flattened cone, similar to No. vj of the French. Cupped. Used by the infantry..................	....	vij	....	xv
3. Large pistol ball, used by the cavalry. Round..................	....	vj	ij	xiij
4. "Ball a stilo." A large conical ball, with a solid base and three rings	i	v	....	i
RUSSIAN.				
1. Long conical ball, with a very deep cup in the base. Three shallow rings. Range 1200 yards...........	i	vj	....	....
2. Liège rifle-ball. Conical, flat base. Three rings and two raised ribs to fit grooves in the barrel...............	i	v	....	iij
3. Conical ball, with flat base, and no rings for the grooved rifle.......	i	vj	....	....
4. Flattened cone with a cup...........	i	i	ij	xiv
5. Same, only a smaller size...........	i	i	....	x
6. Same, only a smaller size...........	i	....	....	xxv
7. Round musket-ball....................	i	....	....	vj
Sometimes two of these round balls were united by wire, so as to resemble bar-shot, and produced great destruction of the soft parts.				

Description of ball.	Average weight.			
TURKISH.	Ounces.	Drachms	Scruples	Grains.
1. Round ball for musket with flint-lock	....	v	ij	....
2. Do. percussion	....	vj	i	iv
3. Conical, flat base, three rings for rifle	i	iv	....	vj
4. Small round. Used by the Bashi Bazouks..............................	....	v	....	....

The **Russians** employed ten sizes of grape-shot, varying in weight **from** oz. i, ℨiv, ℈ij, **and gr. xij, up to** the enormous " whopper," **as the soldiers termed them, whose weight** was lb. iij, ℨij, **and gr. vj. The greater number of the sizes** intermediate between **these two were** either **inclosed in canister** or in bags of **canvas,** or fastened round a spindle of wood. The largest size, above alluded to, was screwed between segments of wood containing hollows fitted to receive the shot. The effect produced by this grape was little inferior to that caused by round shot.

31*

APPENDIX G.

TABLE No. 1.—*Showing the Mortality following the Greater Amputations in all parts for Gunshot Wounds.*

	Primary. No. of cases.	Deaths.	Mortality per cent.	Secondary. No. of cases.	Deaths.	Mortality per cent.	Total cases.	Total deaths.	Total ratio mortality per cent.
British army, Crimea, after April 1, 1855	440	163	37·0	60	36	60·0	500	199	39·8
British Naval Brigade	45	14	31·0	18	9	50·0	63	23	36·5
French, Constantinople—Monnier	156	50	32·0	102	42	41·0	258	92	35·6
French, Constantinople, Sept. 1, 1854, to April 1, 1855—Legouest	8	7	87·0	35	18	51·0	43	25	58·0
Toulouse, 1814—Guthrie	48	10	20·8	51	22	43·0	99	32	32·3
New Orleans, "	45	7	15·5	7	5	71·4	52	12	23·0
Waterloo*	147	61	41·4	225	130	57·7	372	191	51·3
Siege of Antwerp—Larry (fils)	54	9	16·6	10	5	50·0	64	14	21·8
Gros-caillou, 1830 "	6	3	50·0	11	6	54·5	17	9	52·9
Le Hôtel-Dieu, 1830—Menière	15	9	60·0	9	8	88·0	24	17	70·8
Roux, 1830	10	3	30·0	4	4	100·0	14	7	50·0
Spain—Alcock	36	22	61·0	41	14	34·0	77	36	46·7
Communicated to Academy in 1848	29	13	44·8	15	9	60·0	44	22	50·0
Baudens, 1848	8	3	37·0	6	6	100·0	14	9	64·0
Navarino—Sigmare	31	1	3·21	8	2	25·0	39	3	7·6
Indian campaigns†							105	28	26·6
Danish army, 1848–50—Djörup							243	96	39·5
Campagne Constantine—Sédillot							81	17	21·0
Feroe	60	2	3·3				60	2	3·3
27th and 29th Brumaire—Larry (père)	13	2	15·3				13	2	15·3
Battle of Newbourg—Percy	92	6	6·5				92	6	6·5
Faure				300	270	90·0	300	270	90·0
Abonkir and Camperdown	30	0	0·1				30	0	0·0
Communicated to the Academy in 1848, in which no distinction is made between primary and secondary operations							76	34	44·7
St. Louis, 1832—Richerand	15	11	73·3				15	11	73·3
Total	1288	396	30·7	902	586	64·9	2687	1157	43·8

If, however, we calculate only those numbers which furnish the distinction between primary and secondary operations, and exclude such extravagant figures as those given by Signore; if, in short, we take the numbers afforded by the fourteen results which stand first on the table, then the totals will stand as follows: **Primary** operations, 104·; deaths, 374, or 35·7 per cent. mortality; secondary operations, 594, deaths, 314; ratio mortality, 52·8 per cent.; and the average of mortality from both classes of operations combined is **41·9** per cent.

TABLE No. 2.—Showing the Mortality following the Greater Amputations for Injuries.

	Primary.	Deaths.	Mortality per cent.	Secondary.	Deaths.	Mortality per cent.	Total cases.	Total deaths.	Total ratio mortality per cent.
Malgaigne	20	8	40·0	5	1	33·3	173	111	43·9
St. Thomas's Hospital—South	27	8	29·0	26	11	42·0	23	9	39·9
University College Hospital—Erichsen	47	37		45	26	58·0	38	19	35·8
Glasgow Infirmary—Lawrie	75	62	41·0	56	35	65·3	131	63	52·0
,, ,, Steele	169	66	39·6	87	61	70·0	225	99	43·9
,, ,, M'Ghie	180	9	33·3	6	1	16·6	267	121	45·0
Radcliffe Hospital—Hussey	50		18·0				64	10	17·8
Phillips							613	313	50·0
Lyons, 1834—Laroche	18	13	68·0				19	13	68·0
Exeter—Faunz	68	18	29·4				68	18	26·4
Massachusetts Hospital—Hayward									33·3
Total	608	216	35·3	224	187	61·0	1618	776	47·9

* I have in this estimate supposed one-third of these "remaining" in the table furnished by Mr. Guthrie to have died. This computation is a very moderate one, considering the very large number of cases, the results of which were not indicated at the time his table was constructed.

† These numbers include only those cases the results of which I have been able carefully to trace. However successful they may appear to have been, they are nothing to a series reported by Inspector-General Burke, and referred to in the appendix to the first volume of Annesley's Works. Of 80 amputations performed at the siege of Bhurtpore, all recovered within a fortnight!

‡ Besides the primary and secondary operations reported by Signore, he had also 29 intermediary ones with 10 deaths.

APPENDIX II.

Since the completion of the foregoing volume, a work has appeared by M. Scrive, inspector-general of the French medical service in the Crimea, from which a few details are added, as bearing on the medical history of the late war.

"In four months," says M. Scrive, "47,800 men, in an effective force of 145,000, entered our ambulances during the winter of 1855–56; 9000 of these died; an equal number of those transferred, perhaps, perished in the hospitals at Constantinople, or in France."

TABLE *showing the cases and deaths from wounds and disease in the French army from September,* 1854, *to July,* 1856.

	Admissions	Discharged.	Transferred	Dead.
Officers wounded, ordinary.	135	104	31	....
" " by gunshot.	1,625	740	770	115
" ill of fever	1,098	401	503	194[1]
Soldiers wounded, ordinary.	5,582	3,168	2,154	260[2]
" " by gunshot.	35,912	10,178	22,121	3,613[3]
" " frost-bite.	5,596	2,012	3,472	112
Intermittent fever............	6,983	3,746	3,197	40[4]
Remittent fever	12,267	4,036	6,436	1,795[5]
Pernicious fever...............	275	73	52	156[6]
Typhoid fever..................	6,351	1,060	1,628	3,663[7]
Typhus fever	11,124	1,266	3,840	6,018
Diarrhœa........................	19,339	5,248	12,115	1,984[8]
Dysentery.......................	6,105	1,252	2,792	2,061[9]
Cholera	12,258	3,049	3,196	6,013
Scurvy..........................	23,365	4,550	17,576	639[10]
Different fevers	42,453	6,902	34,420	1,731
Venereal	1,455	1,201	241	13[11]
Skin affections, itch	1,255	1,128	124	3[12]
Total	193,178	50,106	114,668	28,404

[1] Many died of typhus and cholera.   [2] Complicated with typhus.
[3] Complicated with cholera, typhus, and scurvy.   [4] By complication.
[5] Frequent complication with typhus.   [6] Cholera and typhus complications.   [7] Many cases of typhus.   [8] Cholera and scurvy complications.   [9] Ditto.   [10] Typhus complications.   [11] By complication.
[12] Ditto.

Cholera prevailed most in June, 1855, and least in March of the same year. In the former month 5466 cases, and 2733 deaths, appear, and in the latter only 3 cases, and 1 death. It was never, however, totally absent up to the end of 1855, when the return closes.

Typhus yielded 711 cases, and 329 deaths, in an effective force of 90,000, during the first winter, and 10,413 cases, and 5689 deaths, in a total force of 140,000 men, during the second winter.

Scurvy was at its maximum in February, 1856. In January, July, August, September, and December, 1855, it was also very prevalent, but by June, 1856, it had greatly diminished. In February, 1856, 4341 cases, and 156 deaths from scurvy, appear in the returns.

Remittent fever prevailed most in August, 1855, and least in June, 1856. Frost-bite gave 3520 cases to the hospitals, followed by 28 deaths, during the first winter, and 2076 cases and 84 deaths during the second.

Table of wounds.

Rank.	Treated in the ambulances.	Discharged cured.	Sent to Constantinople.	Died in the Crimea	Killed.	Slight wounds, treated with their regiments
Officers...............	1,625	740	770	115	325	
Non commissioned officers and soldiers...............	35,912	10,178	22,121	3613	7182	} 1500
Total.........	37,537	10,918	22,891	3728	7507	

Total loss, adding the wounded who did not enter the hospitals, and deducting 3500 Russians, 43,044 men killed and wounded.

Sixteen thousand were killed, or died of their wounds, or after operation. The taking of the city cost the French 5000 wounded, besides the many killed. In twenty-two months 114,668 sick or wounded were transferred from the Crimean to the Constantinople hospitals. The average was 5733 a month and 190 a day, but in August and June it rose to 350 a day, and in the

latter month a total of 10,600 were thus dispatched to the Bosphorus. The proportion of very serious wounds, as compared
with merely severe or slight injuries, averaged 1 in $2\frac{4}{5}$. Nearly
the half had their lives placed in jeopardy by their wounds received in the trenches; 1 in 5 died on the place of combat, and
the same proportion was presented at the battles of the Alma,
Inkerman, and Traktir. The average gravity in these latter
affairs was 1 in 4. Finally, primary amputation was performed
in the proportion of 2 in 12 wounded.

TABLE *showing the degree of gravity from wounds and the accidents of the
field, and the general number of deaths.*

Degree of gravity.	Total cases.	Propor-tion.	Total deaths in the Crimea and Constanti-nople.	Propor-tion.
Very severe—fatal on the field........................	7,507	1 in 5·7	7,507	1 in 1·5
Do. calling, or not. for the removal of a limb........	13,936	1 in 3·1	5,513	1 in 2
Medium severity............	8,317	1 in 5·1	2,300	1 in 3·3
Slight........................	13,284	1 in 3·3	1,000	1 in 13 8
Total..............	43,044		16,320	

TABLE *showing the number of amputations* **and** *resections performed* **in** *the Crimea, and their results.*

Part.	No. of cases.	Proportion of each kind to general total.	Died in the Crimea.		Transferred to Constantinople, or cured.	
			No.	Average.	No.	Average.
Hip...................	12	1 in 397	1275	1 in 3·7	3423	1 in 1·3
Thigh................	1512	1 in 3·1				
Knee.................	58	1 in 80·5				
Leg	915	1 in 5·1				
Foot.................	241	1 in 19·5				
Toes.................	220	1 in 21				
Shoulder...........	168	1 in 28·2				
Arm.................	912	1 in 5·1				
Elbow, forearm, or wrist..........	278	1 in 17				
Hand and fingers	282	1 in 17				
Lower jaw (resection and trephine............	100	1 in 41·6				
Totals.......	4698		1275		3423	

The following shows the proportion which wounds of different regions of the body bore to one another:—

Head....................................	1 wound in	3·4
Neck....................................	1 "	46
Thorax................................	1 "	12
Abdomen.............................	1 "	15
Sup. extrem.........................	1 "	6·2
Infer. "	1 "	4·3

These figures confirm the observations made in the body of the book on the frequency of wounds of the head and thorax in siege operations. The following is the order of frequency of these wounds received in an open engagement:—

Head....................................	1 wound in	10
Neck....................................	1 "	12
Thorax................................	1 "	20
Abdomen.............................	1 "	40
Sup. extrem.........................	1 "	4·3
Infer. "	1 "	3·5

TABLE *showing the wounds and diseases treated in the hospitals at Con-stantinople during the war.*

	Transferred from Varna or the Crimea.	Admissions at Constantinople.	Dismissed cured or well.	Transferred.	Died.
Wounds (ordinary)	2,185	1,007	2,059	720	413
" by gunshot	22,891	.....	9,616	8,190	5,085
Frost-bite...........	3,472	142	2,009	775	830
Typhus..............	3,840	4,889	3,544	1,778	3,407
Cholera.............	3,196	2,570	2,529	1,076	2,161
Scurvy.............	17,576	3,851	9,587	8,460	3,380
Fever	63,124	8,038	35,625	22,988	12,549
Venereal............	241	2,597	2,316	522	......
Skin affections (itch)..............	124	156	256	24	......
	16,649	23,250	6–7,541	44,533	27,825
	139,899			139,899	

To these add the 12,000 or 13,000 who died at Varna, Gallipoli, etc., and the total loss by death is found to be 63,000, viz.: in the Crimea, 28,404; Gallipoli, Varna, Piræus, etc., 5500; Constantinople, 27,823; in the Turkish hospitals, 12,000 to 13,000.

The French had in Turkey fifteen hospitals, containing in all 10,850 to 11,850 beds.

M. Scrive shows most conclusively how much more severe the duty was, which, toward the end of the siege, devolved on the French, as compared with the English army. He gives ample proof of the good effects of chloroform, and states that no disagreeable results whatever arose in their army from its use, and he gives his entire adhesion to the necessity for early amputation in all cases except those at the hip-joint, when he thinks delayed operations do best. M. Scrive's volume contains much interesting and valuable information, chiefly on the diseases which committed such ravages among the French troops in the Crimea.

APPENDIX I.

Since the foregoing pages were printed, the Government Report on the Medical History of the War has appeared. The following *resumé* is added, along with the Surgical Statistics, not given in the text.

The medical part of the report not having as yet been issued, no extracts are given from it. However, such *medical* statistics as are required to elucidate the *surgery* of the war have been obtained through the kindness of the Director-General, and have been given in the body of the book.

It may be here remarked, that the following numbers refer to the period after April 1st, 1855, so far as the *men* are concerned, and to the whole war for the *officers*, except when otherwise stated.

The total number killed in action during the whole war amounted to 2598 men, 157 officers, or 2·7 per cent. of the total force of men sent out, and 4·0 per cent. of the total strength of officers.

The proportion of the various classes of wounds to each other, and to the whole number of wounded per cent., and the mortality by each per cent. of cases treated among the officers for the entire war, and the men for the second period, is shown in the following table :—

	Proportion per cent. of total wounded.		Mortality per cent. of cases treated.	
	In men.	In officers.	In men.	In officers.
Gunshot wounds of the head	11·9	8·1	20·0	17·0
" " face	7·4	6·9	2·6	...
" " neck	1·7	3·2	3·1	10·5
" " chest	5·8	9·3	28·1	31·5
" " abdomen	3·2	5·7	55·7	51·5
" " perineum and genitals	0·7	0·7	30·9	...
" " back and spine	4·5	5·0	13·8	10·3
" " upper extremity	30·2	19·0	2·9	3·6
" " lower extremity	31·7	35·5	8·3	7·2
Sword and lance wounds	0·1	0·5	...	...
Bayonet wounds	0·5	1·7	11·1	...
Miscellaneous wounds and injuries	1·7	3·8	4·7	4·5
Of the above injuries, the following proportion required amputation or resection.	10·8	7·7	27·8	33·3

The following table shows the distribution of the cases:—

Arm of the Service.	Non-commissioned officers and privates.					Commissioned officers.				
	Total sent out.	Killed in action.		Died in hospital of wounds and injuries.		Total sent out.	Died in hospital of wounds and injuries.		Killed in action.	
		Number.	Ratio per cent. of total sent out.	Number.	Ratio per cent. of total sent out.		Number.	Ratio per cent. of total sent out.	Number.	Ratio per cent. of total sent out.
Cavalry..................	8,293	114	1·3	33	0·4	427	8	1·9	4	0·9
Artillery...............	10,723	121	1·1	63	0·6	388	10	2·5	1	0·3
Sappers and Engineers....	1,644	32	2·0	23	1·4	95	9	9·4	6	6·3
Foot Guards.............	6,504	} 2331	3·1	} 1642	2·2	225 }	180*	4·3*	75*	2·5*
Infantry of Line........	66,795					2770 }				
Total..................	93,959	2598	2·7	1761	1·8	3305	157	4·0	86	2·2

* Guards, Line, and Staff.

No. 1.—*Return of wounds and injuries received in action, admitted for treatment in the hospitals in the Crimea, from the first landing at Old Fort to the end of March, 1855. (Non-commissioned officers and privates only.)*

	Total received for treatment.	Died in the field hospitals.	Discharged to duty from the field hospitals.	Discharged to be readmitted under the head amputation or excision.	Transferred to Scutari.	Remained under treatment, on 31st March, 1855.
Gunshot wounds of the head.	206	25	94	...	84	3
" " face	125	...	54	...	66	5
" " neck	64	1	14	...	44	5
" " chest	153	32	14	...	104	3
" " abdomen.	100	36	8	...	53	3
" " perineum and genito-urinary organs.	15	4	...	...	11	...
Gunshot wounds of the back and spine	84	13	20	...	49	2
Gunshot wounds of the extremities	1763	38	251	252	1176	46
Sword and lance wounds	77	...	22	...	55	...
Bayonet wounds	30	3	2	...	20	5
Miscellaneous wounds and injuries.	2	...	1	...	1	...
Particulars not known, no records of them having been kept, or the records having been lost	1815	132	46	4	1633	...
Total wounds	4434	284	526	256	3296	72
From the above cases, 256 primary amputations and resections are returned, which were thus disposed of	...	22	6*	...	220	8
	4434	306	532	...	3516	80

* Of portions of fingers.

No. 2.—*Showing the final result of 4434 wounds received during the above period.*

	Transferred from the Crimea for further treatment.	Died on the passage from the Crimea to Scutari.	Received at the secondary hospitals for treatment.	Died in the secondary hospitals.	Discharged to duty from the secondary hospitals.	Invalided to England.
Gunshot wounds...............		68	3001	332	1234	1435
Sword, lance, and bayonet wounds......................	3296	...	146	5	129	12
Gunshot contusions and miscellaneous wounds....			81	10	35	36
Primary amputations...........	216	14	202	41	...	161
Primary resections.............	4	...	4	1	...	3
Final result of 4434 wounds..	3516	82	3434	389	1398	1647

"It will be seen by return No. 1 that the particulars of 1815 cases are not reported. Either they were not recorded, or in the confusion of the battles of Alma, Balaklava, or Inkerman, the records descriptive of the nature of the wounds received were lost, and all that can now be determined is, that the above numbers were wounded, and died of their wounds before arrival at the secondary hospitals on the Bosphorus.

"It is also to be regretted that the form of return adopted at Scutari, during this period, does not admit the results of the various wounds to be shown by regions, as in return No. 1. This is, however, the less of moment for statistical purposes, as the particulars of so large a number as 1815, or nearly one-third of the total, being unknown, would render any numerical inferences drawn from this series of wounds very imperfect. To the series of the second period, or that following March, 1855, this remark is not applicable, as the returns were kept with the greatest care, and it is believed the results may be implicitly relied on."

No. 3 —*Return of Wounds and Injuries received in action.* (*Non-commissioned officers and privates only.*) *Showing the numbers and results of cases treated from the 1st of April, 1855, to the end of the war.*

	Total treated.	Died.	Discharged to duty.	Discharged to be re-admitted under the local amputation or excision.	Invalided or transferred.
Gunshot wounds of the head............	851	179	594	....	87
" " face	533	14	445	....	74
" " neck............	128	4	108	....	16
" " chest	420	118	226	....	76
" " abdomen	235	131	71	....	33
" " perineum and genito-urinary organs	55	17	23	....	15
Gunshot wounds of the back and spine.	326	45	225	....	56
" " extremities.......	4436	254	2526	766	890
Sword and lance wounds..................	7	1	5	....	1
Bayonet wounds...........................	36	4	22	....	10
Miscellaneous wounds and injuries......	126	6	94	....	26
Total of wounds treated. (72 of which remained on March 31, 1855,).......	7153	764	4339	766	1284
From the above cases 766 amputations and resections resulted, 5 of which required further operation, and 8 cases of amputation remained on 31st of March. Total, 779, which were thus disposed of	....	217	*27	5	530
Add the 8 amputations remaining on 31st March	8	......	......	......	......
Final result of the total number treated during the above period	7161	981	4366	771	1814

* Of portions of fingers.

The foregoing and following returns include gunshot and other injuries analogous to those received in action, such as wounds by the accidental discharge of fire-arms, and injuries received at the great explosion of reserve ammunition on the 15th November, 1855. Of the total number of men "invalided or transferred," viz., 1814, only 1671 were invalided to England; the remainder went to duty from Scutari. A large proportion of officers "invalided," returned to duty subsequently to **their** arrival in England.

No. 6.—*Return of Wounds and Injuries received in action* (*Commissioned officers only.*) *Showing the number and results of cases treated from commencement to the end of the war.*

	Total treated.	Died.	Discharged to duty.	Discharged to be re-admitted under the head amputation or excision.	Invalided.
Gunshot wounds of the head............	47	8	29	....	10
" " face	40	......	29	....	11
" " neck..........	19	2	8	....	9
" " chest..........	54	17	12	....	25
" " abdomen......	33	17	9	....	7
" " perineum and genito-urinary organs.................	4	......	1	....	3
Gunshot wounds of the back and spine.	29	3	18	....	8
" " extremities	318	19	122	45	132
Sword and lance wounds................	3	......	3	....	......
Bayonet wounds........................	10	......	7	....	3
Miscellaneous wounds and injuries. ...	22	1	4	....	17
Total of wounds treated............	579	67	242	45	225
From the above cases 45 operations resulted, which were thus disposed of...................................	......	15	......	....	30
Final result of total number treated...	579	82	242	....	255

The general result, then, is that—

1. Of wounded non-commissioned officers and privates received for treatment during the first period, viz., 4434, 777 died, or 17·5 **per** cent.; 1930 returned to duty, or 43·5 per cent.; 1647 were invalided to England, or 37·1 per cent.; and 80 remained under treatment in the Crimea on 31st of March, 1855, carried into the second period.

2. Of wounded non-commissioned officers and privates received for treatment during the second period. viz., 7081, (and 80 remained under treatment in the Crimea on 31st March, 1855,—total 7161,) 981 died, or 13·7 per cent.; 4509 returned to duty, or 63·0 per cent.; 1671 were invalided to England, or 23·3 per cent.

3. Of wounded officers treated during the entire war, **viz.**, 579, 82 died, or 14·1 per cent.; 242 returned to duty, or 41·7 per cent.; 255 were invalided to England, or 44 per cent.

Making a grand total of 12,094 wounded officers and men treated, (exclusive of 2755 killed in action, as given in returns furnished by the adjutant-general,) of whom 1840 died, 6681 returned to duty, and 3573 were invalided home.

GUNSHOT WOUNDS **OF THE HEAD.**

Besides the numbers given in chapter vii., and which refer to **the** men only, the following cases occurred among officers during **the** whole war. Contusions, more or less severe, 38, with no **dea**ths; contusion or fracture, but no apparent depression, 2, with 1 death; same, with depression, 2, and 2 deaths; 5 penetrating wounds, and 5 deaths.

The remarks made in **the** body of the book, on the management of these cases, **are** fully confirmed by the writer of the report. He states, **moreover,** that "symptoms of compression from abscess of **the** brain" **usually** came on from the fifteenth to the 30th day. A **form** of injury is also spoken of in the report which I have never **seen,** viz., the cavity of the skull being **opened** by the knocking off of the mastoid process.

In 23 cases **an** operation for **the** removal or elevation of depressed bone **was** performed, and 7 recovered. The following gives some of the leading points **of interest** connected with them:—

1. Three slight lacerated wounds on the left side of the head; insensibility for some time after injury, but apparent **recovery**—with headache and febrile symptoms for a day **or two.** Symptoms set in a month afterward, (intermediate treatment not given;) violent pain in his head, drowsiness and stupidity, distortion of face, febrile symptoms, puffy swelling at wounds. Incision showed separation of pericranium; depression overcome by Hey's saw and elevator; dura mater inflamed, but untorn. Symptoms relieved, but hernia appeared. Final recovery.

2. Lacerated wound over right parietal, and comminuted fracture, with considerable depression. Headache, but no febrile symptoms. Ten days after injury, cord-like pain in the head, and hearing affected. Leeches, and general bleeding to twenty ounces, without relief. Delirium. Right pupil dilated. **Pulse 100, and hard.** Skin hot and dry. Trephined on the eleventh **day.** External and internal tables separated, and a piece of wood impacted in the fracture and resting on dura mater. Two circles removed by trephine, and more bone by Hey's saw, to permit of rectifying the injury. Dura mater uninjured. Immediate relief and final recovery.

3. **Fissured** fracture, with slight depression, by shell wound. On the fifth day, **slight symptoms of compression, with** febrile **state. On exposing the bone,** some of the hair was **found fixed in the fracture. Depressed bone** removed by trephine. Inner **table more injured than external, part** being detached. **Blood on dura mater, which was entire. Recovery.**

4. Primary operation. Recovery.

5. Compound fracture of squamous portion of left temporal. Bleeding from the ear. Four large pieces of bone removed, and another elevated. Dura mater entire. **From** being **insensible on** admission, he slowly recovered during the following days. Deafness complete on left side, and partial on right. Suppuration from the ear. Had headache, and a stupid, vacant expression of face when sent to England.

6. Operation three days after injury. Recovery.

7. Keefe, whose case is fully given at page 192.

These were the 7 successful cases of interference. The following were fatal :—

1. Fracture of the right parietal. No symptoms for thirteen days. Rigors and drowsiness then set in, high pulse, pain in head, and intolerance of light; 20 leeches applied. Purgative and calomel (gr. ij) every two hours. Convulsive spasms, and finally paralysis of arm and dilatation of pupils. Operation on the 14th day. Inner table fractured and depressed. Clot on dura mater, which was entire. Slight relief at first, but ultimate death on the twenty-second day. Abscess found in right hemisphere, and fungus growth on surface of the dura mater.

2. Depressed fracture of posterior part of right parietal. Three days afterward dilatation of right pupil, and headache,

double vision, fever, etc. Had leeches to temples, and been bled previously. Trephine applied. Small part of dura mater sloughy, and covered with lymph. Small quantity of pus between dura mater and bone. Considerable bleeding from vessels of scalp. Calomel (gr. ij) every three hours, and antimonials every two hours. A week after operation worse, and bleeding to thirty ounces ordered, with calomel and antimony repeatedly. Jaundice before death by coma, on the eleventh day from operation, and sixteenth from injury. Membranes of brain found inflamed, and abscess in the posterior lobe extending to place of injury.

3. Compound depressed fracture of the left temporal bone. Some loose fragments removed four days afterward. On the fourteenth day, jaundice and delirium, and seven days afterward death. Treatment not stated. " A piece of the internal table was found in the cavity of an abscess, the size of a walnut, at **the seat of** injury."

4. Bone, to a small extent, denuded by a shell wound. **Memory impaired on the** sixteenth day; he was also stupid; had, in succession, pain in the head, dilated pupils, and rigors. On incision of the scalp only a very small scale of bone was found loose, and was removed. Next day pus and air escaped at each beat of the heart by the parieto-frontal suture, which had been laid bare by the explorative incision. Head symptoms increased. Trephined. Part of inner table found depressed and detached. Dura mater inflamed, but entire. Two days after, jaundice and fever, and death on seventh day from operation, and twenty-fourth from injury. Large abscess found in center of right hemisphere, communicating with the wound, and the surface of the hemisphere was coated with pus.

5. Compound comminuted and depressed fracture of anterior inferior angle of right parietal bone by shell. Headache alone for three days. Journey from camp to Castle hospital exhausted him much. On fourth day (next after journey) symptoms of compression. Trephined, and much bone, which was depressed and broken, removed. Dura mater lacerated, and lymph effused on the brain. Sensibility returned when the bone was elevated, and he improved much. For two days only headache and intolerance of light. He then felt as if "something gave way in his head," and a hernia appeared. He got worse, and finally sank comatose, on the fifth day after operation, and ninth from injury.

Dura mater found lacerated and sloughy; brain below place of injury softened, and a large abscess in the right anterior lobe of the brain. Pus over whole external surface of right hemisphere.

6. Depressed fracture of the middle of the frontal bone by shell. Headache and constitutional disturbance. Trephined on the sixth day, and depressed and detached bone elevated. **Dura** mater slightly lacerated. Lost power of utterance twelve days after operation, and, notwithstanding active treatment, got rapidly worse. Convulsions, paralysis of the sphincters and bladder, coma and death on the sixteenth day from operation, and twenty-second from injury. Previous to death a trocar was inserted, to the depth of an inch, into the brain, in the disappointed hope of evacuating pus. Lymph was found deposited in the arachnoid, and much serum in the ventricles. "It seems worth noting that the symptoms in this case, which seemed to have resulted from a nearly pure attack of arachnitis, are almost, if not quite, identical with those observed in many cases where abscess of the brain was the main post-mortem lesion found."

7. Shell wound, denuding small part of frontal bone. Very little complaint for a month, when he was seized with headache, vertigo, and intolerance of light. Pupils and pulse little affected. Treatment for a fortnight by calomel, purgatives, tartar emetic, and low diet. No bleeding. He got worse—became almost idiotic. On the fifty-third day from injury, and the twenty-fourth from the setting in of bad symptoms, an exploration was made, which discovered a piece of the outer table loose, and on this being removed the inner table was seen to be fractured and depressed. Two inches of bone were removed by the trephine. A small fragment was found driven through the dura mater. This was removed. Calomel was given freely. Repeated large bleedings. Got better for a day, and ultimately grew worse, and death relieved him eighty-eight days from the time of injury, and four days after the application of the trephine. Lymph found on the surface of the dura mater at the seat of injury. A sharp piece of bone was found impacted in the brain, and from it, nearly to the base of the anterior lobe, an elongated abscess, distinctly circumscribed, was found filled with sanio-purulent fluid. Pus over the surface of the right hemisphere, and clotted blood near the wound. The arachnoid was inflamed. The inner table fissured beyond what was removed, and new bone was plentifully deposited at the seat of fracture.

8. Ball partially impacted a little above and behind the right ear. No urgent symptoms, except headache, for twenty-three days, when he was sent to the rear. Next day after journey headache increased, and other head symptoms followed. Ball then easily removed, but as the hemorrhage which succeeded caused much faintness, deeply depressed portions of bone were not interfered with. Relief followed the immediate removal of the ball, but his symptoms having progressed, the trephine was applied next day. The dura mater was extensively lacerated. Improvement for two days, and then hernia cerebri, and profuse purulent discharge. Hernia treated by slicing, caustic, and pressure. Flow of pus suddenly ceased, and fungus rapidly increased by the twenty-fifth day after the operation. Became rapidly comatose. By means of a grooved director much pus was **evacuated** during the next two days, and he again became sensible. He continued so for three days, the hernial protrusion having, in the mean time, sloughed away. **He** again became comatose, notwithstanding the free escape of pus, and he died two months after the receipt of the wound, and thirty-four days after the operation. Nearly the whole right hemisphere, as far as the ventricle, was found converted into an abscess with a distinct lining. Pus was found in the other ventricles, **and lymph** abundantly **effused at the base and** other parts. Another and distinct abscess was found at the base of the middle lobe of the left hemisphere.

9. Depressed fracture of occipital bone, with round ball impacted, which was removed by the trephine three days after its introduction. The withdrawal of a small spiculum **of** bone, which had penetrated the torcular herophili, was followed by easily-arrested venous hemorrhage. The dura mater was intact. He **had no** bad symptoms for twelve days, when he was seized with **what** appeared to be congestion of the lungs, and died in a week, the wound being then nearly healed. Nothing except signs **of** recovery were found in the head.

10. Round ball lodged deep in temporal muscle, **with extensive** fracture of the bone. No symptoms except headache for **three** days, when signs of compression appeared. **By** the trephine twelve fragments of bone and a clot of blood were removed, but he never rallied, and died in thirty-six hours. The treatment from the first had been strictly antiphlogistic; but this patient

was a very bad subject, being plethoric and intemperate. No post-mortem.

11. Depressed fracture of right temporal and parietal bones, followed by well-marked signs of compression. Trephined. Dura mater torn, and a slight escape of cerebral matter. Became immediately sensible, and went on improving till the formation of a large hernia cerebri took place. Forty-one days after injury headache and stupor set in, followed by paralysis of the left side, and death fifty-three days after wound and operation. Dura mater found adherent all round the wound, and pus at the base and in the right ventricle.

12. Compound depressed fracture of the skull. No symptoms for a fortnight, then those of compression. Trephined. No relief to symptoms. Death. Membranes much inflamed, and effusion of blood into the brain.

13. Fracture with depression. Primary operation. No relief; and death, comatose, 5 days after operation, from encephalic inflammation, softening of the brain, and abscess.

14. Compound depressed fracture of the frontal bone. Operation demanded by symptoms within 24 hours. No relief. Death 20 hours after operation. Found to be fracture of the base, with effusion.

15. Compound depressed fracture of occipital. Operation, hernia, and death.

16. Compound depressed fracture of left temporal bone. Operation, and speedy death.

17. Operation on appearance of symptoms "some days" after injury. Depressed fracture of occipital. No relief. Death. Fracture extended to the basilar process, and "one edge of the internal table had been slightly raised along the whole of fissured fracture."

18. Extensive depressed fracture by shell. Operation with Hey's saw. Did well for 20 days, then signs of compression, and death from abscess.

19. Another similar case, only living two days.

It is to be regretted that the treatment pursued during the period which intervened between the receipt of the injury and the appearance of dangerous symptoms has not been recorded in the interesting notes given of the above case. The following, then, show the total result of operative interference in these cases :—

Cases treated. Total.	Recoveries without operation.	Recoveries after operation.	Total recoveries.	Deaths without operation.	Deaths after operation.	Total deaths.
76	14	7	21	36	19	55

All the cases which had undergone operation are reported from Chatham to suffer still from vertigo and headache.

There were discharged from Chatham, 88 men, on account of disabilities arising from head injuries. These men suffered chiefly from headache, partial paralysis, mental weakness, or affections of the special senses.

GUNSHOT WOUNDS OF THE FACE.

" Five hundred and thirty-three cases came under treatment among the men during the second period, or 7·4 per cent. of the entire wounded. Fourteen of the patients died, or 2·6 per cent. of those treated. One of these deaths was caused by tetanus, under **which** head the case has been given, the eyeball having been destroyed and the optic nerve injured ; two by inflammation of the membranes of the brain supervening, where one eye had been destroyed ; and two by the same cause, where no injury to this organ had taken place, but the bones of the face had been extensively injured. In four, **very extensive and** deep injury and laceration of the face, **including the tongue, had been i**nflicted ; **two were** complicated with extensive burns from explosion, and in the remaining three the cause of death is not specially reported. Among the officers no fatal case occurred."

The following case is well worthy of being extracted :—

" Private Robert Cuthbert, aged 19, 31st Regiment, was **wounded** on 2d September **by a** grape-shot, which struck **him in the** face, badly fracturing **the** lower jaw. **On removing the bandages** which had been placed on the parts in the trenches, the fractured bone, with its muscles, glands, etc., fell down on the cheek, dragging the tongue with it, and exposing the interior of the mouth and throat as far as the root of the tongue, and the wound **extended** into the anterior triangle of the neck, exposing the caro**tid artery.** The bone was so comminuted, that no choice **was left** but to remove the fragments, and the jagged ends of the bone were sawn even on each side. No part anterior to the angles of the bone could be saved : the soft parts were then brought together, and retained by sutures, and a few adhesive strips and wet lint applied. The patient was now able to lie down, which he could not do before, as the tongue, by falling back, closed the glottis; but even now, when in the recumbent position, he had frequently to lay hold of the tongue, and draw it forward, to facilitate breathing. A considerable portion of the injured integuments of the chin sloughed away, but by careful feeding, dressing, and bandaging, the deformity was ultimately much less than could have been expected. In this case, of course, the food was required to be in the liquid or semi-liquid state, and for a long time great difficulty was experienced in feeding him; but he experienced much comfort from the use of a small pipe, through which he sucked his food. He was sent to England on 24th November, well, and much good might have been expected to result from an operation in remedying a portion of the deformity, as soon as the parts were sufficiently consolidated to warrant such a proceeding."

In the 33d Regiment, the external carotid was successfully tied for primary hemorrhage from it.

In one case, a comrade's double tooth was found imbedded in a patient's eye; in another, tetanus arose from injury to the optic nerve; and in a third, a portion of another man's skull was removed from between the lids of a patient.

GUNSHOT WOUNDS OF THE CHEST.

Return of officers during the entire war.

	Total treated.	Total died.	Discharged for duty.	Invalided or transferred.
1. Simple flesh contusions and } Slight............ wounds..................... } Severe............	11 14	... ...	7 5	4 9
2. With injury of bone or cartilage, but without known lesion of the contents, and not opening the cavity..	6	1	...	5
3. With lesion of the contents, but not opening cavity.....................................	1	1	...	...
4. Penetrating the cavity, and missile lodged, or apparently lodged.............................	8	7	...	1
5. Perforating, or apparently } Superficially...... perforating the cavity... } Deeply..............	1 13	... 8	... ...	1 5
Total...	54	17	12	25

"Of these wounds, 420 have been treated among the men, being 5·8 per cent. of the entire wounded during the period. Of these men, 109 died in the primary hospitals; 8 in the secondary hospitals; and 1 of intercurrent disease, (idiopathic fever,)—making a total of 118 deaths, or 28 per cent. of the cases treated. The very large proportion of the deaths in the primary hospitals is remarkable, and arose from the great severity of the injuries received, by which death was in very many instances caused within a few hours, and in this and the following class especially an unusually large proportion of these fatal cases, under more ordinary circumstances, would have been mortal on the field, and therefore only admitted on the returns as 'killed in action.' From the proximity of the trenches, however, they were rapidly transferred to the hospitals, where, although the care and skill of the surgeon might perhaps prolong life for a few hours, it was but too evident that many were hopeless from the commencement.

"Among the officers 54 cases occurred, or very nearly 10 per cent. of the whole wounded, of which number 17 were fatal, being 31·6 per cent. of the cases treated, a proportion in both instances considerably over that among the men."

In one case of contusion of the walls of the chest, an abscess formed between the pleura and the parietes.

GUNSHOT WOUNDS OF THE ABDOMEN.

Return showing the nature and results of cases treated from the commencement to the end of the war. (Commissioned officers only.)

	Total treated.	Total died.	Discharged to duty.	Invalided.
1. Simple flesh contusions and wounds — Slight	8	....	8	....
Severe	6	....	1	5
2 and 3. Penetrating, or apparently penetrating or perforating the cavity, with lesion — Nature not accurately known	2	1	....	1
Of viscera	15	14	....	1
5. Fracture of the pelvis, not being at the same time wounds opening the cavity of the abdomen	2	2	....	....
Total	33	17	9	7

"For these injuries, including gunshot fractures of the pelvic bones, there were treated 235 men during the period, and 33 officers during the entire war.

"As might almost *a priori* have been expected, this class presents by far the highest rate of mortality of any of the regional wounds: 55·7 per cent of the cases treated having proved fatal among the men, and 51·5 among the officers. In fact, where penetration of the abdominal cavity by gunshot injury was considered to be beyond doubt, death was the rule, recovery the rare exception—only nine patients (including both men and officers) having survived out of 120 where this was believed to have taken place, and even of this small number some of the cases were not quite unequivocal."

GUNSHOT WOUNDS OF THE PERINEUM AND GENITO-URINARY ORGANS, NOT BEING WOUNDS OF THE ABDOMEN OR PELVIS.

Return showing the number and results of cases treated from 1st April, 1855, to the end of the war. (Non-commissioned officers and privates only.)

Total treated	55
Died ⎧ In the regimental or primary hospitals	16
⎨ In the secondary hospitals	1
⎩ Of other disease while under treatment for wound	0
Total died	17
Discharged to duty	23
Invalided	15

Return showing the number and results of cases treated from the commencement to the end of the war. (Commissioned officers only.)

Total treated	4
Total died	0
Discharged to duty	1
Invalided	3

GUNSHOT WOUNDS OF THE BACK AND SPINE.

"The limits of this class are not very clearly defined. In this series, it has been made to include all lesions of the spinal cord; also, lesions of the vertebral column, unless they were, at the same time, wounds penetrating the chest or abdomen, while the flesh wounds and contusions have been, as far as possible, confined to those of the muscles of the spinal column, strictly so called: 326 of these injuries have come under treatment, being

APPENDIX. **389**

4·6 per cent. of the total wounded during the period. Of these patients 45 died, or 13·8 per cent. of the cases treated; 225 returned to duty, or 69 per cent., while the remainder were invalided. Among the officers, 29 cases occurred, of which 3, or 10·4 **per** cent., were fatal.

"All the fractures of the vertebræ were promptly fatal, except two among the officers and two among the men, all of which **were** either fractures of the transverse processes in the neck, or of the spinous processes only. Even where the spinal cord, apparently, **was not** primarily injured, inflammation of it or its membranes was sometimes set up, and quickly proved fatal.

"The functions of the spinal cord were occasionally destroyed temporarily or even permanently, where no discoverable lesion existed, probably in somewhat **the same way as** concussion of the brain produces insensibility."

GUNSHOT WOUNDS **OF THE** EXTREMITIES.

Besides the figures given in the body of the book, the following table refers to the officers during the whole war, which numbers **are** not included in those previously given :—

Return showing the number and results of cases **treated from the commencement to** *the end* **of the war.** *(Commissioned officers only.)*

	Total treated.	Total died.	Discharged to duty.	Discharged and re-admitted under the head amputation or resection.	Invalided.
1. With direct injury of the larger arteries. not being at the same time cases of compound fracture..........	1	1	...	...	...
2. With direct injury of the larger nerves, not being at the same time cases of compound fracture......................	1	1	...	...	...
3. With direct penetration or perforation of the larger joints......................	10	3	1	5	1
4. Of upper extremity, not included above	106	4	41	20	41
5. Of lower extremity, not included above	198	10	80	18	90
Particulars not reported....................	2	...	...	2	...
	318	19	122	45	2

"The instances where wounds of the arteries have been sufficiently distinct and uncomplicated to warrant their being kept separate and returned under this head, have been very rare, only two such being returned during the first period, twelve in the period now under consideration, and one in an officer."

In one post-mortem examination of a case in which death took place after amputation for gangrene resulting from wound of the foot, the following interesting condition was found :—

"The ball was found to have passed through the thigh internally to the sheath of the femoral vessels, which it had grazed but not opened. The artery at this point was slightly contracted for about an inch in length, but pervious, and contained no coagulum, and beyond the contraction its caliber showed no marks of inflammation. The vein, however, was not only also slightly contracted, but its internal surface was inflamed and filled with partially-organized lymph, as far up as the entrance of the deep iliac vein, and downward for about two inches from the wound. Its course was thus entirely sealed, but nothing like pus could be found in the femoral or iliac veins, nor in the venous system anywhere."

GUNSHOT WOUNDS, WITH DIRECT INJURY OF THE LARGER NERVES, NOT BEING AT THE SAME TIME CASES OF COMPOUND FRACTURE.

Return showing the number and results of the cases treated from 1st April, 1855, to the end of the war. (Non-commissioned officers and privates only.)

	Total treated.	Died.			Total died.	Discharged to duty.	Discharged and readmitted under the head amputation or resection.	Invalided.
		In regimental or primary hospitals.	In secondary hospitals.	Of other disease while under treatment for wounds.				
1. Lesion of brachial or axillary plexus	5	1	1	....	2	1	....	2
2. Lesion of median nerve	6	....	....	....	....	....	....	6
3. Lesion of ulnar nerve	4	....	1	....	1	1	....	2
4. Lesion of sciatic nerve	5	3	....	....	3	....	....	2
5. Nerve not specified	2	2	....	....	2	....	....	...
Total	22	....	....	....	8	2	....	12

" Only 22 such are returned among the men, and one in an **offi**cer, of which eight of the men and the officer died, being 41 per cent. of the cases. There can be no question that many flesh wounds occurred, in which nerves of considerable magnitude were more or less implicated; but as they were followed by no special evil consequences, they do not appear to have been returned as injuries to nerves. Of the 9 deaths reported, 5 took place from tetanus — of these some account has been already furnished — 2 from extensive injury to upper part of the thigh, with lesion of the sciatic nerve, and in 2 the cause of death is not specified."

GUNSHOT WOUNDS WITH DIRECT PENETRATION OR PERFORATION **OF** THE LARGER JOINTS.

Return showing the number and results of the cases treated from the 1st of April, 1855, to the end of the war. (Non-commissioned officers and privates only.)

	Total treated.	Died.		Of other disease while under treatment for wounds.	Total died.	Discharged to duty.	Discharged and readmitted under the head amputation or resection.	Invalided.
		In regimental or primary hospitals.	In secondary hospitals.					
1. Shoulder-joint	17	2	.....	1	3	.....	14	.....
2. Elbow-joint...............	30	4	.....	....	4	.....	20	6
3. Hip-joint...................	10	3	.....	....	3	.....	7	.....
4. Knee-joint.................	23	3	.....	....	3	.....	13	7
5. Ankle-joint	8	.....	1	....	1	.....	6	1
6. Joint not specified........	33	11	.....	....	11	.....	19	3
Total	121	.....			25	.....	79	17

*Return showing the number and results of cases treated from the commencement to the end **of** the war (Commissioned officers only.)*

	Total treated.	Total died.	Discharged to duty.	Discharged and readmitted under the head amputation or resection.	Invalided.
1. Elbow-joint......4	0	1	3	0
2. Knee-joint 6	3	0	2	1

"Of 30 injuries to the elbow among the men, and 4 among the officers, 4 were fatal without operation. 2 of these were complicated with injury of the artery, (1 of the branchial, and 1 of the ulnar,) and the fatal result seems to have been mainly due to the combined effects of shock and loss of blood. In 1, extensive, although individually unimportant injuries coexisted—and the cause of death is not recorded in the fourth. 16 primary resections of the joint, or of portions of it, were performed, and 4 secondary ones; while 6 men were invalided, having recovered without operation, with a varying amount of stiffness or partial anchylosis of the joint. In these last, however, there had often been but little injury to the bone inflicted; thus of the officers, one belonging to the 68th Regiment returned to duty, notwithstanding that the injury was followed by a degree of stiffness of the joint; but there was some amount of doubt as to whether it had been primarily opened, although the ulna in its immediate neighborhood was undoubtedly injured. In the remaining instances among the officers, the degree of injury was such as to demand amputation.

"Wounds of the wrist-joint have been returned under the head of injury to the carpus.

"Ten cases of wound of the hip-joint are returned. In 3, as before mentioned, there had been such extensive injury inflicted that they proved fatal in a few hours; 7 were 'discharged for operation;' one of which was for amputation at the hip-joint, in a case in the 34th Regiment, of extensive longitudinal fracture into the joint; the remaining 6 for resection of the head of the bone.

"Of wounds of the knee-joint, 23 cases are returned among the men, and 6 in officers; 6 of these patients died—viz., 3 men and 3 officers. In one of these fatal cases, in the 44th Regiment, both knee-joints were involved, and death took place from shock in twelve hours; in another, in the 68th Regiment, it occurred on the tenth day, the result apparently of inflammation and suppuration of the joint; and in the third from the same cause, at a little later period. There can, however, be no question but that a large proportion of the 33 cases and 11 deaths, in which the specific joint involved has not been reported, were of this description. The cause of death in the three officers is not specially reported."

"The results, then, as above given, do not appear very encouraging toward an attempt to save the limb when any of the larger joints (the knee more especially) is involved, as we have seen that the recoveries in the last-mentioned injuries, without operation, amounted to one-third only of the cases reported. With regard, however, to those of the upper extremity, no great harm appears to have resulted from the attempt at preservation, secondary operations having proved available without any large addition to the risk of life by the proceeding. Of the knee, however, such cannot be said, secondary amputation of the thigh having proved very fatal. The amount of injury done to the bone appears to have been a most important element in determining the treatment of such cases. In none of the recoveries from gunshot wounds of this joint does the bone within the capsule appear to have been more than grazed, (not fractured;) indeed, small fissures into the joint often rendered secondary amputation necessary, or proved directly fatal, as in the case of a man of the 71st Regiment, accidentally shot in the street of Balaklava by a small revolver bullet. The missile had imbedded itself in the tibia just below its tuberosity, whence it was easily turned out by a pointed instrument after a small incision had laid the site open. The knee-joint did not appear to have been involved, but the man died eight days afterward, from the effects of acute inflammation of it, and the accompanying sympathetic fever. On examination after death, a minute fissure was found to have extended through the head of the tibia into the joint. The constitution and previous habits of the patient also appear to have been of much importance."

Besides the numbers given in chapter x. of gunshot wounds of the extremities, and which numbers referred merely to the period subsequent to April 1, 1855, the following occurred among officers during the entire war :—

Upper extremity.—Flesh contusions, and wounds more or less severe, 59, with 3 deaths; simple fracture of long bones, by contusion of round shot or shell, 3, no deaths; contusion and partial fracture of long bones, 2, and no deaths; compound fracture of the humerus, 17, 1 died and **7 were** amputated; compound fracture of forearm, 5, 4 of which were amputated; compound fracture of radius or ulna alone, 2. There were 7 cases of penetrating or perforating wounds of the carpus, 4 of which were amputated or resected.

Lower extremity.—Flesh wounds, 145 cases, and 4 deaths, and 1 submitted to operation; simple fracture by contusion, 2; partial fracture, 2; compound fracture of femur, 20, with 5 deaths and 10 operations; compound fracture of tibia and fibula, 11, 1 death and 5 operations; compound fracture of tibia or fibula separately, 3, with operation. There were 14 cases of penetrating or perforating wounds of the tarsus, and 1 operation in consequence.

" Wounds of the extremities have been tolerably equally divided between the upper and lower limbs, the preponderance having been slightly in those of the latter; the resulting mortality, however, has been very unequal, that of injuries of the upper extremity having been only 2·2 per cent., (exclusive of amputations and resections,) while that of the lower reached 7·5 per cent.; and had the results of the operations performed been added, the difference would have been increased in a very material degree.

" Epiphyses were sometimes knocked off, in the upper extremity chiefly; as happened to an officer of the royal artillery at the great explosion on the 15th November, 1855, in whom the epiphysis on the internal condyle of the humerus was thus taken away, but the elbow-joint not opened; and a man of the 55th Regiment was struck on the tip of the olecranon by a musket-ball. A small portion of the bone exfoliated, but the joint remained perfect, and he returned to duty after four and a half months' treatment.

" The success attending conservative attempts was, to a certain degree, encouraging in the upper extremity, but in the lower sadly the reverse. The percentage of recoveries in gunshot fractures of the several bones (neither death having ensued, nor amputation having been resorted to either as a primary or secondary proceeding) is as follows: Humerus, 26·6; forearm, both bones, 35·0; radius only, 70·0; ulna only, 70·0; femur, 8·0; leg, both bones, 18·8; tibia only, 36·3; fibula only, 40·9."

As much as 3 inches in length of the entire thickness of the humerus was removed in one instance, and the patient recovered, with a useful arm.

"Attempts at preservation of the limb in fractures of the radius and ulna were even more successful than in the case of the humerus, as the percentage of recoveries will already have made apparent; and it is remarkable that where one only was broken, the ratio was the same for each of these bones, and exactly double that existing when both bones had been fractured. Excision of portions of the radius and ulna, although not perhaps, on the whole, so successful, in respect to the perfect use of the limb retained by the patient, as those of the humerus, were also resorted to, and with good effect, both as primary and secondary **operations.**

"In **gunshot** compound fractures of the femur, the deaths returned in the table at page 103 only amount to 82 per cent. of the cases treated without operation, and among these a large proportion of limbs torn off by round shot are included, which, as we have already seen, was almost always a fatal accident, while the percentage of deaths in amputation **of** the thigh reaches 65·2. From this it would at first sight appear that the success attending attempts at the preservation of the limb with fracture of this bone had **been** nearly as great as that in those cases where **the** limb was condemned **and removed. It must**, however, be borne **in mind that** the recoveries without operation have only amounted to 14 out of 174 cases among the men, and 5 out of 20 among the officers, or 10 per cent., and that all those selected for the **experiment** of preserving the **limb were so chosen** expressly on account of the comparatively **small** amount of injury done both **to the** bone and the soft parts, and that even then recovery was always tedious, and the risks during a long course of treatment numerous and grave. In many cases, also, amputation of the thigh was performed because death was otherwise evidently inevitable, and it was thought right the patient **should be** allowed the benefit of the chance, however small, afforded by operation. **By** this means the number of deaths **among the** operations is swelled, while, at the same time, the number of deaths due to the fracture is **by so much** diminished, and the percentage, therefore, tells doubly in favor of treatment without operation.

"On account of the very indifferent success of amputations **of the** thigh, a trial was made of resection of portions **of** the shaft of the bone; but no success attended the experiment, every case, without exception, **having proved** fatal where this was attempted."

There were only **7 sword and** lance wounds received after April 1, 1855, and 1 death therefrom: 2 were of the head, and 1 died; 1 of the face, 1 of the perineum, 1 of the back, and 2 of the

lower extremities. During the same period, 36 bayonet wounds
were received, and four **died therefrom**: 1 was of the head, and
was fatal; one of the face; 11 of the chest, 2 of which were
mortal; 2 were of the abdomen; 2 of the back; 3 of the **upper**
and 13 of the lower extremity; 1 had numerous **such wounds**;
and in 2 the site was not specified. Among the officers **during**
the war, 3 sword and lance wounds, 2 bayonet wounds of the
abdomen, and 8 of the lower extremity are reported, with no fatal
result therefrom.

The following table gives **the** number of cases of amputation
and resection in officers during the entire war. These numbers
were not included in those given in chapter xii., which referred
merely to the men.

*Return showing the number and results of cases treated from the com-
mencement to the end of the war. (Commissioned officers only.)*

			Total treated.	Total died.	Invalided.
Amputa-tions of	Upper extremity..	Shoulder-joint........	6	2	4
		Arm	7	1	6
		Forearm	4	......	4
		Thumbs	2	1	1
		Fingers...............	6	......	6
	Lower extremity..	Hip-joint..............	2	2	......
		Thigh { upper third.	5	4	1
		middle "	2	1	1
		lower "	5	3	2
		Leg	5	1	4
	Double operation....	One leg and one foot.	1	......	1
			45	15	30

"The comparative ratio of mortality in the several operations,
between the men and the officers, is slightly greater among the
former; thus **the ratio per** cent. of cases of the amputation
treated is as follows:—

	Men.	Officers.
Double operations	50·0	......
Toes	......	*
Medio tarsus	14·3	*
Ankle	16·6	*
Legs	35·6	20·0
Knee-joint	55·5	*
Thigh, **lower third**	56·6	60·0
Thigh, middle third	60·0	50·0
Thigh, upper third	87·1	80·0
Hip-joint	100·0	100·0
Fingers and thumbs, etc	0·9	12·5†
Forearm	5·0	......
Arm	26·4	**11·3**
Shoulder joint	**33·4**	**33·3**

"It is difficult to assign a reason **for** this difference with any degree of certainty."

"And the following **will** show the results of the operations received for treatment, **and** performed at Scutari during the portion of the first period named therein, beyond which time it is regretted that this distinctive table was not kept up; **but** it **is** known that very few operations indeed **were** performed there after the end of November.

"The **ratio** of mortality shown in it from primary operations **appears very** disproportionately small, from the number received **for** treatment having been reduced, first by deaths in the **field** hospitals, and secondly by deaths on board ship, during **the** passage to the Bosphorus; the particulars of which, **for the indi**vidual operations, as before stated, cannot now be **arrived at, but** the rate of mortality, upon the gross total of **primary** operations performed, appears slightly to have exceeded 50 per cent., showing that what was advanced in the first section of this report, as the general opinion **among the** army medical officers, as to the cases of the first period having suffered severely **by carriage in** the Crimea and on **the** voyage to the secondary **hospitals, was** well founded, as primary operations in **the second** period only

---

* **None done.**

† **One death** took place under peculiar circumstances of constitution; and **as** the number of cases **is** small among the officers, **the** rate of mortality is thus unduly increased.

show a mortality of 35 per cent., (operations on fingers and toes being excluded in both.) The ratio in secondary amputations appears to have been on the whole nearly the same as obtained in the series of the second period, as recorded on the preceding page."

Return of the number and results of amputations and resections treated in the hospitals on the Bosphorus from the 26th September to the 27th of November, 1854.

			Primary.					Secondary.				
			Total treated.	Died.	Remained under treatment.	Invalided to England.	Ratio of mortality per cent.	Total treated.	Died.	Remained under treatment.	Invalided to England.	Ratio of mortality per cent.
Amputations of	Upper extremity.	Shoulder-joint...	6	1	2	3	16·6	.....	.....	.....	.....	.....
		Arm.............	44	2	17	25	4·5	10	3	3	4	30·0
		Forearm..........	14	.....	3	11	.....	7	2	1	4	28·7
		Hand at wrist....	2	.....	1	1	.....	1	.....	.....	1	.....
	Lower extremity.	Thigh	44	8	11	25	18·2	33	27	1	5	81·8
		Leg.............	35	5	6	24	14·3	13	9	2	2	69·2
		Ankle-joint.......	1	.....	.....	1	.....	.....	.....	.....	.....	.....
		Medio tarsus.....	2	1	.....	1	50·0	.....	.....	.....	.....	.....
		Tarso metatarsus.............	1	.....	.....	1	.....	.....	.....	.....	.....	.....
	Double operations.	Both arms	1	.....	.....	1	.....	.....	.....	.....	.....	.....
Resections of		Head of humerus	2	1	.....	1	50·0	1	1	.....	.....	100
		Elbow-joint......	2	.....	.....	2	.....	.....	.....	.....	.....	.....
			154	18	40	96	11·6	65	42	7	16	64·6

It seems that amputation at the hip was 14 times performed in our army, in place of 10 times, as stated in the text; and as the result was always fatal, this goes still further to strengthen the deductions made in chapter xii. The patient who survived longest only lived 36 hours.

"In concluding this report on the surgical practice of the late war, it may not be improper to notice that the number of wounds and injuries treated, as set forth in the General Return A of Sick and Wounded, exceeds very considerably that herein given; the former (among non-commissioned officers and privates) being 18,283; the latter 11,515 only. This apparent discrepancy arises from the general return embracing all admissions for mechanical injuries, including kicks from horses, accidental cuts and bruises, and the innumerable minor accidents to which the soldier, in common with all working men, is exposed; by which he, like them, is occasionally for a time disabled, and of which it has not been thought necessary in this report to enter into a detailed account; while the smaller number is confined to wounds and injuries, either actually received in action with the enemy, or strictly analogous in nature."

"Something under 6 per cent. of the strength then would seem to indicate the limit beyond which reserved hospital accommodation need not be kept for the reception of the wounded of a large army engaged in active field operations, while it is equally plain that much under 5 per cent. would not be safe.

" It is scarcely necessary, however, to observe that the proportion of wounded in any individual member of the component parts of a lager force may be very widely different from that here stated; thus, at the battle of Inkerman, the 41st and 95th Regiments, with a strength in the Crimea of 678 and 500 respectively, received into hospital for treatment 104 and 120 cases of wounds, or 15·3 per cent. of the strength in the former, and 24·0 in the latter, and even these numbers appear to have been exceeded in some corps on other occasions.

" It is also of some importance, as bearing upon the number of recruits necessary to be sent out to keep an army in the field at a given strength, to ascertain with accuracy the average number of any given series of men disabled by wounds received in action who return to duty as effective soldiers, and the average time they remain under treatment before this result is obtained.

" With regard to the first period into which the campaign has been divided in this report, 43·5 per cent. of the men returned to duty. The information on the second of these points, however, is defective, for several reasons, and the time itself was subject to disturbing agencies of various kinds, which did not affect the series of wounds of the second period. During the first, also, a much larger percentage of cases treated was invalided to England, viz., 37·1, against 23·3 in the second. The cause of this was not so much the greater severity of the wounds received, or the less successful treatment, as the pressure on the hospitals during the winter of 1854, which led to the transfer home of all cases fit to be removed, which were likely to require a lengthened period of convalescence before they could be pronounced fully fit to re-

sume the duties of a soldier on active service. **During the latter
period**, as before stated, 7161 wounded **men were received for
treatment**, of whom 4509 returned to duty, or **63 per cent.** The
following table exhibits the time at which this **result** took place
in 6359 of the **cases, of which number** 4015 returned **to** duty.
The information cannot be given for the entire series, **as it has
not been** furnished by **a** few corps, but the proportion known
is so **large that** for practical purposes it seems sufficient :—

	Number treated.	Ratio per cent. returned to duty
Total of wounds treated.............................	6359	.....
Returned to duty after a period of treatment :		
Under one week.....................................	1476	23·2
Over a week, but under one month...........	1408	22·1
Over one month, but under two...............	709	11·1
Over two months, but under three............	263	4·1
Over three months, but under four...........	101	1·6
Over four months, but under five.............	40	0·6
Over five months, but under six..............	11	0·1
Exceeding six months..............................	7	0·1
Total returned to duty..............................	4015	63·1

NOTE.—This table has no reference to men who were invalided to
England, and subsequently returned to duty.

The number and results of the secondary capital operations,
for the effects of wounds received in action, performed in the
general hospitals in this country, (exclusive of the foot guards
and ordnance corps,) will be seen in the following table :—

	Number performed.	Died.
Amputation of arm...	3	....
" forearm..............................	10	....
" thigh.................................	4	1
" leg.....................................	3	....
" toe.....................................	1	....
Removal of diseased bone from stumps........	4	....
Ligature of external iliac artery.................	1	....

The total number of men discharged the service for disabilities
consequent upon wounds received in action, and other mechanical
injuries inflicted during the late war, was 3011. The several
causes are thus shown :—

Disabilities.	Cavalry.	Foot guards.	Regiments of line.	Ordnance corps.	Total.
Luxations.................................	1	....	3	2	6
Gunshot wounds........................	56	187	1755	120	2118
Incised and punctured wounds........	19	....	11	1	31
Contusions.............................	3	2	36	13	54
Fractures..............................	16	1	54	15	86
Burns.................................	...	1	4	...	5
Amputations...........................	11	54	547	59	671
Resections............................	....	....	11	....	11
Injuries not specified.................	4	....	5	20	29
	110	245	2426	230	3011

The proportion **discharged for wounds in the** different regions **was** as follows :—

Gunshot wounds of the head.............................	88
" " face...............................	106
" " **neck**	16
" " **chest**........	71
" " abdomen........................	10
" " perineum, etc.................	8
" " back and spine...............	23
" " upper extremities.............	551
" " lower extremities.............	588
" " lower joints........	66
" " artery.........................	1
" " **nerves**.........................	14
Sword, bayonet, and lance wounds.....................	21
Miscellaneous..	70
Total...1633

To this add 464 cases of amputation **in the upper,** and 179 in **the lower** extremity, also 23 **cases of resection of the** joints of the **upper, and 1 of the head of the femur.**

The account **of the cases of resection of the** head of the humerus and elbow, when they were invalided at Chatham, is not very **encouraging.**

The Alma and the minor affair at the Bulgaria gave 73 officers, and 1536 wounded, and as most of these were from round shot or grape fired at shot range, " the injuries were peculiarly severe, and numerous operations were required." The battle of Balaklava and the affair of the 26th of October gave 36 officers and 329 men wounded. The battle of Inkerman yielded a much larger number still, so that between these affairs and the trench casualties, 4434 wounded non-commissioned officers and privates, with the operations which resulted, are not included in the numbers given in the statistical tables of the war. Of these 4434 cases, 777 died, or 17·5 per cent., while only 981 deaths followed in the 7,161 cases of wounds which occurred subsequently to April 1, 1855, giving a percentage of 13·7.

With regard to chloroform, it appears that that administered to the only patient who died under it was, by the report of Dr. Maclagan, who examined it, " totally unfit for use, being in a state of complete decomposition."

TETANUS.

It appears that in all, 5 cases are alone reported previous to April 1, 1855, while later, 24 cases occurred—thus being only 0·2 per cent. of the wounded.

The following table exhibits a succinct view of the cases treated in the Crimea :—

Regiment.	Name.	Age.	Date of injury.	Nature of wound.	Period at which tetanic symptoms supervened.	Duration of tetanus.	Result.
93	D. Ross	30	18th Aug.	Lesion of sciatic nerve	17th day.	35 days.	Cure
35	R. Swain	...	31 Sept.	Lesion of ulnar nerve	7th day.	2 days.	Death
1 B. R. B.	W. Hardinge	...	8th Sept.	Destruction of eye, and lesion of optic nerve	4th day.	2 days.	Death. {Death on 8th Sept.; a rapid case, but dates not reported.
19	C. Martin	28	8th Sept.	Lesion of axillary plexus	......	......	Death
41	Corp. C. Ventham	...	8th Sept.	Lesion of axillary plexus	12th day.	15 hours.	Death
19	....	...	8th Sept.	Lesion of sciatic nerve	7th day	3 days.	Death
49	J. Leahen	24	8th Aug.	{Shell foot wound—secondary hemorrhage. Amputation of leg on 24th day.	7th day.	4 days.	Death
57	A. Nixon	32	11th Aug.	Primary amputation of leg.	11th day.	......	Death
62	....	...	2d May.	{Primary amputation of thigh, followed by secondary haemorrhage on the 4th day.	aft. amputtn	......	Death
49	....	...	......	Compound fracture of both bones of forearm.	6th day.	4 days.	Death
18	W. Howe	...	18th June	{Primary amputation of arm, and conjoined Fracture of tibia.	8th day. 30th day.	5 days. 4 days.	Death Death
88	Corporal Murphy	...	8th Sept.	{Compound fracture of scapula by bullet.	......	......	Death. **23d Sept.**
19	W. Reek	28	30th Aug.	{Compound fracture of urethana and injury of testicle by grape-shot.	18th day.	4 days.	Death.
2 B. R. B.	P. Donegan	27	8th Sept.	Compound fracture of tibia.	4th day.	3½ days.	Death.
41 B. R. B.	....	...	......	Flesh wound cnap of back.	......	......	Death.
38	....	...	......	Flesh wound perineum.	......	......	Death.
38	....	...	......	Flesh wound lower extremity.	......	......	Death.
44	....	...	......	Flesh wound of leg.	......	......	Death.
Gen. hosp.	B. Hughes	22	18th June	{Flesh wound, situation not reported. Flesh wound of hip and buttock; extending beneath deep fascia: ball extracted by incision 3d day.	......	......	Death 27 days after wound.
Do. 38	J. Barker	19	10th June	Flesh wound thigh; ball extracted by incision.	......	......	Death 17 days after wound.
17	J. Howie	23	7th Feb. 1856.	Frost-bite.	21st day.	39 days.	Cure.
19	M. Rourke	...	......	{Slight abrasion of side of foot, an idiopathic case.	......	......	Death 5th October, 1855.

www.ingramcontent.com/pod-product-compliance
Lightning Source LLC
Chambersburg PA
CBHW021351210326
41599CB00011B/836